Safety Evaluation and Regulation of Chemicals

First International Conference on Safety Evaluation and Regulation of Chemicals,
Boston, Mass., February 24–26, 1982

# Safety Evaluation and Regulation of Chemicals

Editor
*F. Homburger,* Cambridge, Mass.

18 figures and 64 tables, 1983

KARGER

S. Karger · Basel · München · Paris · London · New York · Tokyo · Sydney

National Library of Medicine, Cataloging in Publication
International Conference on Safety Evaluation and Regulation of Chemicals (1st: 1982: Boston, Mass.)
Safety evaluation and regulation of chemicals/First International Conference on Safety Evaluation and Regulation of Chemicals, Boston, Mass., February 24–26, 1982
Editor, F. Homburger. – Basel; New York: Karger, 1983
1. Consumer Product Safety – congresses 2. Drug Screening – methods – congresses 3. Toxicology – standards – congresses I. Homburger, F. (Freddy), 1916– II. Title
W3 IN1898M 1st 1982s
[QV 771 I589 1982s]
ISBN 3-8055-3578-3

Drug Dosage

The authors and the publisher have exerted every effort to ensure that drug selection and dosage set forth in this text are in accord with current recommendations and practice at the time of publication. However, in view of ongoing research, changes in government regulations, and the constant flow of information relating to drug therapy and drug reactions, the reader is urged to check the package insert for each drug for any change in indications 7nd dosage and for added warnings and precautions. This is particularly important when the recommended agent is a new and/or infrequently employed drug.


Printed in Switzerland by Thür AG Offsetdruck, Pratteln
ISBN 3-8055-3578-3

# Contents



*State of the Art of Safety Evaluation*



*Risk Assessment*

## *Special Regulatory Problems*



## *In vitro Testing: Mutagenesis, Carcinogenesis*



## *Proffered Papers*



## *Method Development for Safety Evaluation*

## *Alternate Methods for Animal Testing*

# List of Participants

* = Speakers and chairmen

Adams, Richard A., Bio-Research Consultants, Inc., 9 Commercial Avenue, Cambridge, MA 02141, USA

Aldrich, Franklin D., IBM Corporation, PO Box 1900, Boulder, CO 80302, USA

Arthur, Jack, Environmental Protection Agency, 401 M Street SW, Washington, DC 20460, USA

* Ashford, Nicholas, Center for Policy Alternative, Massachusetts Institute of Technology, 1 Amherst Street, Cambridge, MA 02139, USA

Atkins, David, Harvard School of Public Health, 677 Huntington Avenue, Boston MA 02115, USA

* Autrup, Herman, National Institute of Health, 9000 Rockville Pike, Bethesda, MD 20205, USA

* Balazs, Tibor, Food and Drug Administration, 200 C Street SW, Washington, DC 20204, USA

* Baram, Michael, Boston University School of Law, 765 Commonwealth Avenue, Boston, MA 02215, USA

Baron, Ronald, Union Carbide Company, PO Box 12014, Durham, NC 27709, USA

Barth, Delbert S., Biological Science Department, University of Nevada, Las Vegas, NV 89154, USA

Bear, Mark P., Bioassay Systems Corporation, 225 Wildwood Avenue, Woburn, MA 01801, USA

* Bernfeld, Peter, Bio-Research Institute, Inc., 9 Commercial Avenue, Cambridge, MA 02141, USA

Bogle, Gilbert, Environmental Quality Laboratory, California Institute of Technology, Pasadena, CA 91125, USA

* Brown, David R., Northeastern University, College of Pharmacy and Allied Health Professions, 360 Huntington Avenue, Boston, MA 02115, USA

Brown, Halina, Massachusetts Department of Environmental Quality Engineering, 1–11 Winter Street, Boston, MA 02108, USA

* Burns, James P., Jr., Bureau of Drugs, HFD-110, 5600 Fisher Lane, Rockville, MD 20857, USA

Calderbank, A., Imperial Chemical Industries, Ltd., Jealotts Hill Research Station, Bracknell, Berkshire RG12 6EY, England

* Cavender, Finis, The Mitre Corporation, 1820 Dolley Madison Boulevard, McLean, VA 22102, USA

Charbonneau, Sylvia M., Health and Welfare Canada, 204 Health Protection Building, Tunney's Pasture, Ottawa, Ontario K1A 0L2, Canada

Christiano, Rita, Bristol-Myers Products, 1350 Liberty Avenue, Hillside, NJ 07207, USA

* Covello, Vincent, National Science Foundation, Division of Policy, Research & Analysis, 1800 G Street NW, Washington, DC 20006, USA

* Curren, Rodger, Microbiological Associates, 5221 River Road, Bethesda, MD 20816, USA

Davies, Sheila, National Board of Health and Welfare, Department of Drugs, Box 607, S-751 25 Uppsala, Sweden

Dean, Brian J., Shell Toxicology Laboratory, Sittingbourne Research CXENTRE' Sittingbourne, Kent, England

Denine, E.P., Norwich-Eaton Pharmaceuticals, PO Box 191, Norwich, NY 13815, USA

Feltmate, T.E., University of Guelph, Ontario Veterinary College, Department of Veterinary Microbiology and Immunology, Guelph, Ontario N1G 2W1, Canada

Gallant, Raymond, Managing Editor, Food Chemical News, 400 Wyatt Building, 777 14th Street NW, Washington, DC 20005, USA

* Garattini, Silvio, Istituto Ricerche Farmacologiche 'Mario Negri', Via Eritrea 62, I-20157 Milano, Italy

Glomot, R., Centre de Recherches Roussel UCLAF, Départment de Toxicologie, 102 route de Noisy, BP No. 9, F-93230 Romainville, France

* Golberg, Leon, Division of Community Occupational Health, Duke University Medical Center, Box 2914, Durham, NC 27710, USA

Green, Milton, Polaroid Corporation, 730 Main Street, Cambridge, MA 02139, USA

* Grosso, Louis S., Hoechst-Roussel Pharmaceuticals, 1041 Route 201–206 North, Somerville, NJ 08876, USA

* Hayes, John A., Mallory Institute of Pathology, 784 Massachusetts Avenue, Boston, MA 02118, USA

Helms, Donna, G.D. Searle & Co., 4901 Searle Parkway, Skokie, IL 60076, USA

* Henry, Carol J.M., Microbiological Associates, 5221 River Road, Bethesda, MD 20816, USA

Hill, Robert, Syntex, 3401 Hillview Avenue, Palo Alto, CA 94304, USA

Hollebone, J., Pesticide Division, Agriculture Canada, Ottawa, Ontario K1A 0C6, Canada

* Homburger, Freddy, Bio-Research Institute, Inc., 9 Commercial Avenue, Cambridge, MA 02141, USA

* Irving, George W., Jr., Life Science Research Office, Federation of American Societies for Experimental Biology, Bethesda, MD 20815, USA

Jaggers, S.E., Imperial Chemical Industries, PLC, Central Toxicology Laboratory, Alderley Park, Macclesfield, Cheshire SJ10 4TJ, England

Karger, Eva, Polaroid Corporation, 575 Technology Square, Cambridge, MA 02139, USA

Kohn, Fred E., G.D. Searle & Co., 4901 Searle Parkway, Skokie, IL 60076, USA

* Kolbye, Albert C., Food and Drug Administration, 200 C Street SW, Washington, DC 20204, USA

Krebs, Steven, Alcon Laboratories, Corporate Toxicology, Box 1959, Fort Worth, TX 76101, USA

Krueger, Howard W., Procter & Gamble Co., Ivorydale Technical Center, Cincinnati, OH 45217, USA

* Langenbach, Robert, Environmental Protection Agency, Genetic Toxicology Division, Research Triangle Park, NC 27711, USA

* Mazué, Guy, Centre de Recherches Clin Midy, rue du Professeur Blayac, F-34082 Montpellier Cédex, France

Mehlman, Myron A., Mobil Oil Corporation, PO Box 1026, Princeton, NJ 08540, USA

Mekler, Philipp, Sandoz Ltd, Toxicology 881, CH-4133 Pratteln, Switzerland

* Miller, Carolyne T., Health and Welfare Canada, Tunney's Pasture, Ottawa, Ontario K1A 0L2, Canada

* Moustafa, Laila A., International Programme on Chemical Safety, IRRU, World Health Organization, PO Box 12233, NIEHS, Research Triangle Park, NC 27709, USA

* Nichols, Joanne, Center for International Studies, Massachusetts Institute of Technology, Cambridge, MA 02139, USA

* North-Root, Helen, Armour-Dial Company, Armour Research Center, 15101 N Scotts-dale Road, Scottsdale, AZ 85260, USA

* Page, Talbot, Environmental Quality Laboratory, California Institute of Technology, Pasadena, CA 91125, USA

Plank, James B., Westwood Pharmaceuticals, Inc., 468 Dewitt Street, Buffalo, NY 14213, USA

* Prival, Michael, J., Genetic Toxicology Branch, Food and Drug Administration, 200 C Street SW, Washington, DC 20204, USA

* Putman, Donald L., Microbiological Associates, 5221 River Road, Bethesda, MD 20816, USA

* Reznik-Schüller, Hildegard, National Cancer Institute-FCRF, Chemical Carcinogenesis Program, Postbox B, Frederick, MD 21701, USA

Richter, Paul L., 519 Scotland Road, Union, NJ 07083, USA

Rogers, Diana, CRA, Syntex USA, Inc., 3401 Hillview Avenue, Palo Alto, CA 94302, USA

* Schneider, F. Howard, Bioassay Systems Corporation, 225 Wildwood Avenue, Woburn, MA 01801, USA

* Schuler, Ron L., National Institute for Occupational Safety and Health, 4676 Columbia Parkway, Cincinnati, OH 45226, USA

Schuppan, Dietrich, Bundesverband der Pharmazeutischen Industrie, Karl-Strasse 21, D-6000 Frankfurt/Main, FRG

* Scicluna, Henry, Principal Administrative Officer, Council of Europe, BP 431 R6, F-67006 Strasbourg Cédex, France

Segre, Eugene J., Syntex Research, 3401 Hillview Avenue, Palo Alto, CA 94304, USA

Seidman, Floyd, Warner-Lambert Company, 201 Tabor Road, Morris Plains, NJ 07950, USA

* Shirasu, Yasuhiko, Institute of Environmental Toxicology, Suzuki-cho 2-772, Kodeira-shi, Tokyo 187, Japan

Slater, William A., Merck & Co., Inc., PO Box 2000, Rahway, NJ 07065, USA

Smart, C., Mobil Oil Corporation, Toxicology Division, PO Box 1029, Princeton, NJ 08540, USA

Smith, Thomas H.F., IBM, 1 Barker Avenue, White Plains, NY 10601, USA

* Solt, Dennis B., Harvard School of Dental Medicine, 188 Longwood Avenue, Boston, MA 02115, USA

* Somers, E., Environmental Health Directorate, Health and Welfare Canada, Tunney's Pasture, Ottawa, Ontario K1A 0L2, Canada

Stara, Jerry, Environmental Criteria and Assessment Office, Office of Research and Development, Environmental Protection Agency, Cincinnati, OH 45268, USA
Stieghan, Don L., Atlantic Richfield Co., 515 South Flower Street, Los Angeles, CA 90071, USA
* Tardiff, Robert G., Academy of Sciences, 2101 Constitution Avenue NW, Washington, DC 20418, USA
* Todhunter, John, Assistant Administrator, Environmental Protection Agency, Washington, DC 20460, USA
* Ward, Jerrold M., Laboratory of Comparative Carcinogenesis, National Cancer Institute, Frederick, MD 21701, USA
* Weinberg, Edmund, International Flavors and Fragrances, 1515 Highway 36, Union Beach, NJ 07735, USA
* Williams, Gary, Dana Institute, American Health Foundation, Valhalla, NY 10595, USA

# Acknowledgements

The following have assisted in organizing the First International Conference on Safety Evaluation and Regulation of Chemicals, Boston, Mass., 1982:

Local Organizing Committee: *Dean John Sandson,* Honorary Chairman, *Freddy Homburger,* Secretary, *Leonard Gottlieb, John A. Hayes,* and *Edward W. Pelikan.*

International Advisory Committee: *E. Somers,* Health and Welfare Canada, Ottawa (Canada); *I.F.H. Purchase,* Imperial Chemical Industries, Ltd., Alderley Park, Macclesfield, Cheshire (England); *René Truhaut,* Laboratoire de Toxicologie, Paris (France); *B. Schnieders,* Bundesgesundheitsamt, Berlin (Germany); *Marcus A. Klingberg,* Tel Aviv University, Ramat-Aviv, Tel Aviv (Israel); *Yasuhiko Shirasu,* Institute of Environmental Toxicology, Tokyo (Japan); *Gerald Zbinden,* Institut für Toxikologie der ETH und der Universität Zürich, Schwerzenbach (Switzerland); *René Dubos,* Rockefeller University, New York, N.Y.; *Myron A. Mehlman,* Mobil Oil Corporation, Princeton, N.J.; *Sheldon W. Samuels,* AFL-CIO, Washington, D.C.; *Irving J. Selikoff,* Mount Sinai School of Medicine, New York N.Y.; *Raymond E. Shapiro,* NIEHS, Research Triangle Park, N.C. (USA).

The following firms and foundations supported the conference with financial contributions: Allied Chemical Corporation; Arthur D. Little Foundation; Bio-Research Consultants, Inc.; Bio-Research Institute, Inc.; Mobil Oil Corporation; Polaroid Corporation; Sandoz, Inc.; The Sciences and Arts Foundation; Wm. Underwood Company.

Grateful acknowledgement is made for editorial work and various types of technical and organizational assistance in preparing and operating the conference and in editing the proceedings to Mrs. *Helen Bachmann-Bitoun,* editorial assistant.

# Welcome Address

It is with great hope and confidence that I bid you welcome to this International Conference on Safety Evaluation and Regulation of Chemicals, sponsored by the American College of Toxicology and organized by departments of Boston University School of Medicine, the Mallory Institute of Pathology and my own Bio-Research Institute, which thereby marks its 25th year of existence.

I am confident that this will be an interesting and stimulating meeting because of the high quality of our speakers as well as of our select audience.

This conference came about because some time ago the toxicologist of one of our British customers complained that it was difficult for Europeans to be informed about the various regulations and methods for the evaluation of chemicals in various parts of the world and that a conference on this subject would be useful. This stimulated us and we thus established a tentative program with a large group of speakers from many countries. We felt that this meeting should be held at the same time as the meeting of the Society of Toxicology, but, as the date of that meeting was changed, we had to shift our meeting to its present date, with the unexpected result that we now conflict with an international meeting of the German Mutagenesis Society in Berlin, which caused many of our transatlantic participants to drop out in order to go to Berlin.

This is the time when, as secretary of the organizing committee, I have an opportunity to thank the many who have helped to bring this meeting about. We thank especially the speakers who have agreed to come and make this the signal event we hope it will turn out to be. I also thank the audience. Let us hope that their journey and expenses will prove to be worthwhile.

*F. Homburger*

Safety Evaluation and Regulation of Chemicals. 1st Int. Conf., Boston, Mass., 1982,
pp. 1–8 (Karger, Basel 1983)

# State of the Art of Safety Regulation: The Role of the US Environmental Protection Agency

*John Todhunter*

Environmental Protection Agency, Washington, D.C., USA

All countries recognize that due consideration of public health and environmental concerns is part of the responsible utilization of chemicals. Each country, however, approaches these problems from different perspectives. These perspectives are often codified as the statute laws of the sovereign nations.

## *Toxicity – What Is It?*

I have included a short section at this point to outline some various classes of toxicity which are of interest with respect to toxic substance regulation.

Broadly defined, chemical toxicity is the causation of detrimental effects to living organisms by chemical agents. The effects of toxic substances can be immediate (acute toxicity) or long-term (generally a chronic effect) and can be specific to tissues (organ-specific) or generally detrimental (systemic toxicity). Chemical control programs need to address these broad parameters as well as variations within these parameters.

Generally speaking, the following sorts of effects are recognized:

*Acute Toxicity.* Generally a toxic effect resulting from very short-term exposure to a chemical substance. The exposure required to produce an acute effect is generally higher than that which might produce a subchronic or chronic effect. Acute effects may be reversible or irreversible. Of particular concern are life-threatening effects, neurological damage, damage to liver, lungs, kidney or the immune system and acute teratogenic effects.

*Subchronic Toxicity.* Generally a toxic effect resulting from prolonged exposure of limited duration. Exposure levels needed to produce toxicity are often lower than for acute effects. Many of the same concerns listed for acute toxicity are of concern in subchronic exposures.

*Chronic Toxicity.* Generally a toxic effect produced by exposure of a continuous and prolonged nature. Exposure levels involved generally seem to be lower than for effects of a more rapid nature. Delayed toxic reactions, progressive degenerative damage to tissues, reproductive toxicity and cancer are among common elements of this subclass.

*Environmental Toxicity.* This is a catch-all phrase used generally to connote toxic effects upon the environment (fish or bird toxicity for example), or at times to toxicity resulting from exposure to toxicants found in the environment (for example, a toxic reaction to solvents in a paint thinner would not normally be considered 'environmental' toxicity, but a reaction to the same solvents as air or water pollutants would).

Of major importance in the discussion of chemical toxicity is a realization that toxic potential by itself is only part of the risk picture. (In the United States 'risk' is normally taken as the actual or relative potential for harm and 'hazard' generally denotes inherent toxic potential. In many foreign countries, 'hazard' is used in place of and synonymously with 'risk'.) The hazard presented by a chemical can only be manifested if one is exposed to the chemical. Therefore, risk is a function of toxic hazard *and* exposure. Mathematically:

$$R = f(H,E)$$

where R = risk, H = toxic hazard, and E = exposure.

This is really a more general statement of the basic toxicological principle of dose-response which states:

$$r = f(d)$$

or, response (r) is a function of dose (d).

For most toxicological concerns, safety is normally addressed through the establishment of a 'no effect level' (NEL, or the highest dose producing no effects) and then setting a 'margin of safety' (some ratio, usually 100–1,000, sometimes greater depending on the severity of the toxic effect, which is used to set a bound on allowable human exposures).

When the toxic effect of concern is cancer or genetic damage, it has been common to consider that some degree of risk will exist at any exposure level. The theoretical underpinnings of such an approach are currently a matter of discussion within the scientific community, particularly the appropriateness of such an approach for the so-called 'nongenotoxic' carcinogens. Even under the 'no-threshold' concept for setting allowable exposures to carcinogens, it is recognized that the risk levels associated with certain low doses are so low that in practical terms cancers would not be observed in an exposed population. The concept of such a 'virtually safe dose' (VSD) has found considerable utility in the regulation of carcinogens. Most federal agencies tend to use VSD levels that control risks to the range of $10^{-4}$ to $10^{-6}$ or lower. Most agencies tend to use the upper end of the range given as an area where some oversight (of a nonregulatory nature) may be appropriate.

These sorts of considerations lead to the conclusion that one needs to have a two-pronged public/environmental chemical safety program. One prong ought to be prospective: screening new chemicals (those not presently in use) or new uses of chemicals to ensure that these new practices will not result in unreasonable risk. The other prong ought to concentrate on the existing chemical picture to clean up any problems which might currently exist. There are a number of laws dealing with chemicals in the US – e.g., Consumer Product Safety Act, the Federal Insecticide, Fungicide, and Rodenticide Act, Resource Conservation and Recovery Act, and so on. I will focus on the Toxic Substances Control Act (TSCA), because the TSCA best exemplifies this two-pronged approach. I will describe the conceptual framework of TSCA in general – which has a perspective in common with many of the laws of other countries – and then dwell particularly on how we evaluate hazards of new chemicals.

### *The Toxic Substances Control Act*

The TSCA was signed into law by President Ford on October 12, 1976. Upon signing the TSCA, the President stated. '[TSCA] is a strong bill and ... focuses on the most critical environmental problems not covered by existing legislation, while not overburdening either the regulatory agency, the regulated industry, or the American people.'

The TSCA took effect on January 1, 1977. For the past 5 years, the EPA has sought to carry out the TSCA's congressional intent, namely to prevent unreasonable risks of injury to health or the environment associated with the manufacture, processing, distribution in commerce, use or disposal of chemical substances. To support this goal the TSCA's activities center around the following three major areas.

(1) The premanufacture notification (PMN) program provides for scrutiny of the entire life cycle of each new chemical, insuring its safety *before* manufacture or subsequent release into the environment.

(2) The testing of new or existing chemicals can be required when the EPA believes that a chemical may present an unreasonable risk to health or the environment. In order to require testing, the EPA must find that there is insufficient data on the chemical substance and that further testing is necessary before a risk determination can be made.

(3) The control of existing chemicals which are found to pose an unreasonable risk to health or the environment is a third major aspect of the TSCA. Written into the Act itself is a manufacturing, processing, distribution and use ban on polychlorinated biphenyls (PCBs). Certain uses of other chemicals such as chlorofluorocarbons (CFCs), asbestos and dioxin have had limitations placed on specific uses or disposal.

Many of the EPA's evaluations hinge on data gathered from industry under the reporting and record-keeping section of the TSCA. Based on this information, the EPA published an initial Inventory of Chemical Substances in 1979, and in 1980, a revised Inventory. This is the first such inventory ever produced. Any chemical not on the inventory is subject to the PMN program.

In short, the TSCA is unusual among federal toxics-related laws for a combination of reasons. First, it focuses attention on chemicals as products, not as pollutants. Second, its scope includes virtually all chemicals (both new and old), and not just those used for specific purposes. Before the TSCA, the government was primarily concerned with drugs, food additives, and pesticides. Third, it represents a preventive approach to protecting public health and the environment by providing for review of new chemicals before they are introduced to the market. And finally, it is among the few US environmental laws that require the explicit consideration of both risks and benefits before taking action. Under the TSCA, we must determine that the chemical substance either presents or may present an 'unreasonable risk' to human health or the environment. To determine whether a risk is 'unreasonable', the agency must consider the benefits a substance provides as well as the risks it creates.

I would like to turn now to a more detailed discussion of data used and risk assessment under the TSCA. Information on the toxicity hazard of a chemical substance is gathered principally from three sources: epidemiological studies (when available), animal studies, and inference from data on related compounds. Each of these sources has its own limitations in terms of 'certainty' about what the human or environmental hazard actually is.

Next, data on exposure of humans and/or the environment are obtained. Again, a variety of sources contribute data which represent a range of 'certainty'. For example, in some instances we might have specific environmental monitoring data; in others, data on surrogate compounds; and in still others, simply production volume and categories of use information.

In the third stage, the toxicity hazard information is combined with the exposure information in order to assess the risk that the chemical might pose to human health and/or the environment as it moves from manufacture, through use, and then to disposal. However, disposal per se is not generally dealt with through the TSCA.

The various components of risk analysis must then be factored into a determination as to whether or not the risk is 'unreasonable'. This involves consideration of factors such as the benefits which the chemical substance contributes, the existence of and risks associated with the use of alternatives, and the possible health or environmental 'disbenefits' which can result from economic dislocations. Therefore, the TSCA is inherently a risk-benefit balancing statute, a characteristic which distinguishes it from some other pieces of US environmental and public health legislation.

If the risk is judged to be unreasonable, the Act provides for a variety of mechanisms to mitigate the level of risk to one judged to be 'reasonable'. Generally, this involves some degree of exposure limitation; but new or better information on benefits or hazards can also alter the balance of the equation. The Act mandates that the least burdensome of the options which will achieve a favorable balance between risks and benefits be selected.

### *Application to New Chemicals*

I would like to turn now to the 'unreasonable risk' decision, in the case of new chemicals.

First, let me recall that, under Section 5 of the TSCA, anyone who intends to manufacture a new chemical substance must notify the Agency at least 90 days prior to initiating such manufacture; hence the term 'Premanufacturing Notification', better known as PMN. The Act specifies certain information which *must* be a part of the notification: (1) the identity of the chemical; (2) its use(s); (3) the anticipated production volume; (4) a description of byproducts; (5) the number of individuals likely to be exposed in the workplace, and (6) the method of disposal.

In addition, the manufacturer is required to include any test data on the substance relating to its health and environmental effects, and a description of any other such data to the extent that they exist and are reasonably ascertainable by the manufacturer.

Thus, depending on the amount of testing which has been conducted by the manufacturer and the amount of information available in the open scientific literature, a PMN submission could contain anywhere from a great deal to very little toxicity hazard data. In experience to date with the TSCA, few new chemicals are submitted with a great deal of test information. This seems to be largely due to the inherently low hazard potential of the bulk of PMN chemicals. Exposure estimates depend on the production volume, use and worker involvement data – which are all required in the submission.

Often the task of assessing the benefits is even more difficult than that of evaluating the risks. Here, information on production volume, intended uses, and number of workers can be used to develop at least a general picture of what benefits are likely to emerge from introduction of the chemical into the marketplace and what alternatives, if any, exist. The integration of the risk and benefit information provides the basis for a decision on whether or not the proposed new chemical may pose 'unreasonable risks to human health or the environment'.

In the case of new chemicals, the unreasonable risk decision is somewhat more complex than that of existing chemicals. There are really two different types of decisions which can be made under section 5 of the TSCA.

First, under section 5(f) if the administrator, upon review of the information discussed above, determines that there is a reasonable basis to conclude that the manufacture, processing, distribution in commerce, use, or disposal of the new chemical presents or will present an unreasonable risk of injury to health or the environment, action can be taken (a rule or an order issued) designed to reduce exposure to the chemical. Explicit options for a rule include: (1) prohibit specific use(s); (2) limit the concentration of a product; (3) require labeling; (4) require that certain records be kept; (5) control means of disposal, and (6) provide risk information to the public. The proposed order would define and set conditions for the manufacture, processing, or distribution of the chemical in commerce, if such activity is allowable. The thrust of the proposed order would be to mitigate risk to a reasonable level, given a certain level of benefit. In dealing with the roughly 1,000 PMNs received so far, we have not taken a single 5(f) action. This reflects the fact that the vast majority of chemicals which reach the premanufacturing stage have already been effectively screened by industry and are being proposed for uses which, in all likelihood, do not present unreasonable risks.

The second type of unreasonable risk decision under section 5 is one of great interest. Since the TSCA does not authorize the EPA to *require* specific test data for PMN submission, it is possible that there might be some cases when the data are insufficient to reach a decision about unreasonable risk. Therefore, section 5(e) provides that if the data submitted are indeed insufficient to permit a reasoned evaluation of the health and environmental effects *and* that an unreasonable risk of injury to health or the environment *may* be present (or the material is likely to result in significant or substantial human exposure), then the administrator may take action prohibiting or limiting the manufacturing, processing, distribution in commerce, use or disposal of the subject chemical until such time as sufficient data are submitted. To date the agency has taken actions under 5(e) which have banned the production of nine new chemicals, pending the development of specific data judged to be necessary in order to reach a decision concerning 'unreasonable risk'. As an example, in one recent case, consideration of the molecular structure of a substance suggested its possible metabolism in humans to *o*-dianisidine, an animal carcinogen related to benzidine, a known human carcinogen. In addition, the proposed use of the material was such that both workers and segments of the general population might receive significant levels of exposure to the substance. After considering such non-risk factors as the apparent availability of substitutes and the costs of testing and possible control devices, the agency concluded that the chemical might present an unreasonable risk and consequently took action under section 5(e) to gain the additional information upon which to make a firm decision.

The effectiveness of our program should not, however, be measured by the number of 5(e) actions. In fact, most situations in which our concerns are raised never reach the 5(e) stage. Specifically in some 60 cases thus far, the manufacturers have voluntarily agreed to provide the EPA with more data, have taken steps to reduce exposures to the new chemicals at the EPA's request, or have withdrawn the PMNs.

We believe, therefore, that the TSCA provides us with the regulatory tools the EPA needs to avoid unreasonable risks in connection with the introduction of new chemicals.

These, I believe, are the highlights of one highly important and complex law which my office has the responsibility of carrying out. There are, no doubt, questions concerning it which you may seek to have answered. I urge you to remember the story of the brave mountain climber who, having lost his footing, found himself dangling over a sheer cliff with only a small handhold. In near despair, he called out, 'Is anyone there?' A majestic voice replied, 'Yes, my son.' The man asked, 'Will you help me?' To which the voice replied, 'Do you have faith?' When the man answered that he did indeed have faith, the voice instructed him to let go. At this point the man asked, 'Is there anyone else up there that I can talk to?' In short, if you have any questions please call or write my office and I will ensure that they are properly answered. Thank you.

Dr. J. Todhunter, Environmental Protection Agency, Washington, DC 20460 (USA)

Safety Evaluation and Regulation of Chemicals. 1st Int. Conf., Boston, Mass., 1982, pp. 9–15 (Karger, Basel 1983)

# Overview of Safety Regulations Governing Food, Drugs and Cosmetics in the United States

*J. Burns*

Food and Drug Administration, Bureau of Drugs, Rockville, Md., USA

The Food and Drug Administrations's (FDA) mission is to protect consumers by ensuring, among other things, (1) that foods are safe, pure and wholesome; (2) that drugs are safe and effective and manufactured under adequate controls, and (3) that cosmetics are safe. The FDA also regulates other products such as medical devices, animal drugs, biologicals such as blood and blood products, and radiation-producing electronic equipment.

For purposes of this talk, I will confine my comments to safety regulation of food and drugs since cosmetics require no preclearance by the FDA. Although no safety data are required to be submitted, the agency will offer advisory opinions on toxicology studies for cosmetics. The industry is responsible for marketing a safe cosmetic, and the burden of proof is on the FDA to show harmfulness to remove a cosmetic from the market.

The authority and responsibility for the regulation of food and drugs are established by the US Congress. Historically, the first federal authority to act against foods and drugs was the Pure Food and Drug Act of 1906. This Act prohibited interstate commerce in foodstuffs and drugs with false claims of purity and quality. With regard to drugs, the thrust of this original Act was to the control of fraudulent claims for patent medicines and did not include any requirement for the establishment of safety of a chemical prior to its interstate commerce. The weakness of this law was brought into sharp perspective by the 'Elixir of Sulfanilamide' tragedy of 1937 in which more than 100 people, principally pediatric patients, died because a new drug, sulfanilamide, had been dissolved in a poisonous solvent, diethylene glycol. The toxicity of this solvent was never tested in animals prior to marketing because it was not required under the 1906 law. It would have been easily demonstrated to be unsafe by even the most elementary of toxicologic tests.

In response to the public furor over this episode, and realizing the inherent deficiencies of the 1906 Act, Congress passed new legislation in 1938 know as the Federal Food, Drug and Cosmetic Act (FD&C Act). This law, as amended over the years, still stands as the statutory authority for the FDA. This 1938 Act required drug manufacturers, for the first time, to obtain predistribution clearance of a new drug on the basis of safety for its intended use. There was no provision at that time for premarket clearance of food additives. No official notice of drug approval was needed to start marketing a new drug; permission to market was automatic if no word to the contrary was received from the FDA within 60 days after submission of the application (New Drug Application; NDA) to the agency. Although animal safety studies were then required by law, no guidelines existed to achieve any consistency of effort from company to company. Usually, at that time, anything up to a 1-month study in animals was considered acceptable to demonstrate safety. Within 10 years, some manufacturers were conducting studies in rats of up to 3 months duration as a measure of chronic toxicity of a drug. By 1957, 6-month studies in the rat and dog were considered an adequate assessment of the chronic toxicity of a drug. This continues to be the norm in many parts of the world to this day. At the present time, study of the chronic toxicity of a drug to satisfy FDA standards would consist of a 1-year study in a nonrodent and an 18-month study (or longer) in a rodent.

The first appearance of any kind of animal safety study guidelines was in 1949, when a set of criteria for animal safety studies was published in the Food, Drug and Cosmetic Law Journal of that year. Following several revisions, more definitive guidelines were published in 1959. This 'Appraisals Handbook', as it was called, was never advertised as guidelines but they became the standard for animal safety studies in the Food and Drug Industries for some time.

Although the FD&C Act of 1938 already prohibited poisonous or deleterious substances in foods except in certain instances where such substances were allowed in small amounts, the lack of a premarketing clearance requirement had made this provision relatively ineffective. In order to bar the use of a dangerous additive, the FDA had to assume the burden of proof and show the additive was poisonous or deleterious which was a time-consuming process. To correct this, Congress passed the 1958 Food Additives Amendment to the 1938 Act. The major provision of this amendment was the requirement for premarketing clearance of additives which were not generally recognized as safe (GRAS). Among other things, this amendment required data to be provided to the FDA establishing that the additive was safe. Such data were ordinarily obtained from feeding studies and other tests using the proposed additive at various levels in the diet of two or more animal species.

The 1958 Food Additive Amendment contained the controversial Delaney Clause which stated that, '...no additive shall be deemed to be safe if it is found to induce cancer when ingested by man or animal, or if it is found, after tests which are appropriate for the evaluation of the safety of food additives, to induce cancer in man or animal'.

Although the amount of animal toxicity data required to market a drug had increased gradually over the years, it generally lagged behind the requirements of the food safety amendments. As an example, the FDA animal study guidelines from 1949 stated that, since drugs were generally only used for short periods of time, which limited the duration of toxic insult, more emphasis needed to be placed on the development of chronic effects with food additives than with drugs. Within 10 years, however, this distinction was no longer made. The increasing use of potent drugs for chronic diseases and evidence for serious cumulative or latent toxic effects with long-term use of these drugs required more chronic animal safety studies to predict the potential for these chronic toxic effects.

The last major revision of the 1938 FD&C Act with respect to foods and drugs occurred in 1962 with the passage of the Kefauver-Harris Amendments. These were enacted into law following the 'thalidomide disaster' and required drug manufacturers to prove efficacy as well as safety of a drug prior to its approval for marketing (NDA). To accomplish this, the FDA implemented the Investigational New Drug Regulations. At this time, the pharmaceutical industry sought advice of the FDA as to what animal safety studies would be adequate to support the required clinical studies to investigate the efficacy of new drugs. New guidelines for animal safety studies for INDs were first presented in 1963 at a Joint Symposium on the Safety Evaluation of New Drugs. They were first formally announced as guidelines to a congressional subcommittee in 1966 and were subsequently published in summary form in 1968. With some modification, these still form the basis for animal safety studies in support of new drug development. It was not until 1962 that guidelines for the study of the effects of drugs on reproduction and teratogenicity originated. These were a modification of the 3-generation reproduction studies used for food additive safety evaluation and arose as a reaction to the thalidomide tragedy. Although initially the requirement was for a two-litter study, it was soon revised to a three-segment study with each segment covering a specific phase of the reproductive process.

There had been a major interagency effort, mediated by the Interagency

Regulatory Liaison Group (IRLG) to develop consistent guidelines for animal safety testing which would satisfy the safety evaluation needs of all types of chemicals regulated by the various members of the IRLG (CPSC, FDA, EPA, OSHA, and Food Safety & Quality Service of the Agriculture Department). However, the interagency agreement was terminated in September 1981.

Let me now turn to discuss what are the current operating policies with regard to the safety regulation of foods and drugs. Let me first discuss food additives. According to the FD&C Act, a new substance shall be permitted in food if it produces a described technical effect and if it is demonstrated to be safe under the conditions of use. Further, only that amount that adequately serves the purpose should be used, not excessive amounts beyond need. The FD&C Act defines a food additive as 'any substance, the intended use of which results or may reasonably be expected to result, directly or indirectly, in its becoming a component or otherwise affecting the characteristics of any food...'. There is a distinction to be made between the two different types of food additives. There are, (1) Direct Food Additives – those substances intentionally added directly to food and, (2) Indirect Food Additives – unintentional additives or substances that result from contact of food with materials used for packaging, handling, or storing of food. In order to market a food additive, a company must make a formal submission of a Food Additive Petition to the FDA, requesting the agency to issue a regulation providing for safe use of the substance in food as requested.

The type and degree of animal safety testing required in a Food Additive Petition will be largely dependent upon the extent and type of food additive used and the degree of human exposure involved. For a new substance that is intended for deliberate addition to food, e.g. a preservative, an emulsifier, or a thickener, a complete toxicological profile including chronic toxicity/carcinogencity and reproduction studies would be required. At the other extreme would be an unintentional additive that, under any practical condition, would virtually not become a component of food, although at some stage, either in processing or in packaging, had some contact with food. For additives under these conditions of virtually no migration, short-term, acute toxicological data would suffice.

Between these two extremes are various types of studies required, depending on the degree of migration for an indirect food additive and what ancillary information may be available bearing on the safety of that additive. A full safety evaluation of a food additive would include the following: (1) acute toxicity in several species; (2) subchronic studies (90 day) in a nonrodent species to assess cumulative toxicity; if this is to be part of an end-point assessment of an insignificant indirect food additive, a subchronic rodent study of 90 days duration should also be conducted using in utero exposed animals with an assessment of reproductive performance and teratology at the end of the subchronic phase; (3) some assessment of the metabolism of the substance; (4) lifetime studies in two rodent species; one study, usually in the rat, should begin with in utero exposure; (5) reproduction studies in the rat over three filial generations, and (6) a 1-year study in a nonrodent species, provided there are to be two lifetime rodent studies.

Mutagenicity studies are not required at present, but are suggested for screening and guidance prior to long-term studies. Although human testing is not a requirement for approval of a food additive, human toxicology data can come from industrial exposure, poisonings, epidemiology studies and limited controlled clinical experiments. The availability of human data or experience based on common use in food may reduce the animal safety data required for a food additive or may reduce the safety factor. The food additive procedural regulations refer to applying a safety factor of 100-fold based on the 'no-effect' test level in experimental animals. This ordinarily applies when long-term toxicology data are available and a 'no-effect' level can be obtained. In cases of unintentional food additives that can be supported by subchronic toxicity data, a 1,000-fold safety factor is usually applied.

In addition to food additives, the FDA also regulates color additives, i.e., '...a substance added or applied to a food, drug or cosmetic or the human body for purposes of imparting color'. Special animal safety studies directed to the study of the toxicity of a color additive include: (1) dermal irritation/percutaneous toxicity in rabbits of 1–3 months duration; (2) lifetime skin-painting study in mice; (3) sensitization studies in the guinea pig, and (4) sensitization studies in humans with tests for photosensitization. Eye area use of a color additive would also require data from studies by repeated ocular instillation and color additives in sutures would require implantation studies.

Let me now turn to the safety regulation of drugs. As stated earlier, manufacturers of drugs are required to demonstrate the safety and efficacy of a drug in humans prior to approval of a New Drug Application (NDA) for marketing. In order to study the clinical efficacy, appropriate clinical investigations must be conducted. These are done under an Investigational Exemption for a New Drug Application (IND) and require adequate preclinical testing to support such clinical studies. Investigational clinical studies are divided into three phases designed to introduce the drug to an ever-expanding patient population on a controlled systematic basis. The clinical evidence for safety and efficacy developed under the IND will ultimately be used to support approval of the NDA for marketing of the drug. The types of preclinical safety studies needed to support IND clinical studies are largely determined by the phase of clinical investigation that the drug is in, as well as the proposed duration and route of human administration. The animal studies are intended to keep in advance of the phase of clinical investigation.

The three phases of clinical investigation are as follows: phase I – generally represents the first introduction of a drug into man. Clinical studies conducted in this phase are usually of short duration and are directed to studying dose tolerance and the clinical pharmacology of the drug in normal volunteers. Preclinical studies to support phase I clinical studies are usually of 2–4 weeks duration in two species for oral or parenteral drugs. Phase II – clinical investigations consist of controlled clinical trials designed to demonstrate effectiveness and relative safety. Normally, these are performed on closely monitored patients of limited number, about 100–200 patients. The types of preclinical studies recommended to support phase II clinical trials would normally be of 1–3 months duration in two species. Phase III – clinical trials are the expanded controlled and uncontrolled trials conducted to confirm the efficacy of the drug in clinical practice and to determine the clinical safety of the drug. Animal studies to support these clinical trials are normally of 3–6 months duration, but possibly longer.

For ultimate approval of an NDA for marketing of a drug for chronic use, the sponsor would need a 1-year non-rodent toxicity study and a rodent toxicity study of at least 18 months duration, the latter serving as an evaluation of the carcinogenic potential of the drug.

For drugs which will be used to treat women of child-bearing age, reproduction studies will also be needed during the investigational stage. These are divided into three segments as follows: segment 1 – study of fertility and general reproductive performance; segment 2 – teratology studies, and segment 3 – perinatal/postnatal study.

Drugs administered by other than the oral or parenteral route require special safety studies by the proposed route of administration. Certain classes of drugs that are thought to have particular toxicity problems and have the potential for widespread and chronic use, e.g. beta-blocking drugs, also require a more extensive evaluation of the carcinogenic potential, i.e. studies in two rodent species. Combinations of drugs require an assessment of the toxicity of the combination as well as of each component.

Recently, the FDA has issued regulations that have had a significant impact on the safety testing of regulated chemicals. The Good Laboratory Practice Regulations (GLPs) for Non-Clinical Laboratory Studies issued in late 1978 as a result of the discovery by the agency that some studies submitted in support of the safety of regulated products had not been conducted in accord with acceptable scientific practice and that the quality and integrity of these submitted studies had been compromised. It was felt that the seriousness of the problems encountered in testing facilities at the time demanded the need for such GLP regulations rather than informal guidelines. It was argued that decisions about the safety of consumer products that are based wholly or in part on data derived from such testing are too important for the agency to accept anything less than the best scientific data that can be obtained.

In summary, I have attempted in this short time to give an overview of the regulation of the safety of foods, drugs, and cosmetics through the laws, regulations and guidelines of the FDA. It is no easy matter and should involve a continuing dialogue between the industry and the FDA, so that important issues will get a fair and complete discussion. Today's regulatory decisions involve complex choices that are usually not black and white but various shades of grey and are usually close calls. The more input there is into that decision, the better chance that it will be the best decision.

J.P. Burns, PhD, Food and Drug Administration, Bureau of Drugs,
5600 Fishers Lane, Rockville, MD 20852 (USA)

Safety Evaluation and Regulation of Chemicals. 1st Int. Conf., Boston, Mass., 1982, pp. 16–25 (Karger, Basel 1983)

# Regulatory Control of Chemicals: The Canadian Experience

*E. Somers, G.E. MacDonald*

Environmental Health Directorate, Health Protection Branch, National Health and Welfare, Ottawa, Ont., Canada

The extensive and increasing use of chemical substances in industry, agriculture and on public and private properties has brought countless benefits to our society. The introduction of many new chemicals to the Canadian environment and the continued use of many existing substances has, on occasion, proceeded with inadequate knowledge of possible adverse environmental or human health effects. In recent years, society has become increasingly aware of the acute and chronic injury to man and the environment which can result from the improper handling or disposal of many of these substances.

The Consitution Act, 1867, determines specific areas of federal and provincial jurisdiction. Of the powers enumerated in this Act, those associated with criminal law and trade and commerce are the most likely to be used in dealing with the protection of the consumer from risks resulting from hazardous substances.

## *The Criminal Law Power*

There appear to be very few limitations on what the traditional role of the criminal law is to prevent people from engaging in conduct that might be harmful to others, and regulation of the production and distribution of hazardous substances seems to fall fairly within the role.

*The Trade and Commerce Power*

The trade and commerce power is of interest because so many potentially hazardous substances are in the flow of commerce.

The courts have had numerous occasions to consider the scope of the trade and commerce power and it has recently been interpreted more broadly to allow effective federal regulation of interprovincial trade. It may be uncertain whether the trade and commerce power can be relied on to support legislation dealing with hazardous substances, but the criminal law power and the general power appear adequate to allow Parliament to take effective action.

*Provincial Jurisdiction*

Provincial jurisdiction is even broader. The provinces may legislate with respect to manufacturing within the province, with respect to labor relations and the working environment, and with respect to waste disposal. There are, however, some limitations. First, the provinces may not legislate with respect to matters beyond their boundaries. Second, they may not legislate at all with respect to federal Crown property or other classes of subjects that are within exclusive federal jurisdiction. Third, provincial legislation becomes inoperative whenever it conflicts with valid federal legislation.

*Legislative Framework*

Control of toxic substances is required from manufacture to disposal. Legislative control, federal and provincial, can cover the following elements: manufacture, importation, transportation, storage and handling, use or sale, discharge to water, air and land, spills, occupational health, and disposal. There are some 27 federal statutes which exercise some form of control respecting toxic substances. In addition, the 10 Provinces have enacted 99 pieces of legislation.

The relevant federal legislation may be classified as follows:

*General Pollution Statutes*

*The Canada Water Act.* This statute authorizes the establishment of water quality management areas by the federal government in collaboration with provincial governments. Some of the steps to ensure effective management include monitoring, research and the establishment of a water quality plan. The Act also deals with nutrients resulting from the manufacture and use of laundry detergents and other cleaning agents which may cause eutrophication.

*The Clean Air Act.* This Act, like the Canada Water Act, is cast in general terms and implementation depends upon promulgation of specific emission standards. The Act also provides for the federal government to enter into agreements with provincial governments for the purpose of facilitating, coordinating and implementing policies and programs designed for the control and abatement of air pollution and adopting and applying, throughout Canada, national ambient air quality objectives.

*The Fisheries Act.* The Fisheries Act prohibits the deposition of a 'deleterious substance' of any type in water frequented by fish. In addition, the power to make regulations respecting substances, or classes of substances, quantities and concentrations of substances or classes of substances and treatments, processes and changes of water is also included.

*Ocean Dumping Control Act.* This statute has been in force since 1975 and is designed to control dumping of waste and other substances in the ocean. Three schedules have been included within the Act which define prohibited substances, which cannot be dumped (Schedule I); restricted substances, which may only be dumped upon receipt of a permit (Schedule II), and Schedule III, which enumerates factors to be taken into account in granting permits.

*General Contaminants Control Statutes*

*The Environmental Contaminants Act.* This statute is intended to cover problems that cannot be effectively handled under other environmental legislation, particularly 'new' or recently identified pollutants. It is jointly administered by the Ministers of the Environment and National Health and Welfare.

This Act provides the Minister of National Health and Welfare with broad powers to collect data and conduct investigations with respect to substances entering or likely to enter the environment in quantities or concentrations or under conditions which may endanger human health or the environment.

Regulations have been promulgated under this Act to prohibit specified uses of environmental contaminants as follows: polychlorinated biphenyls, mirex, polybrominated biphenyls, polychlorinated terphenyls, chlorofluorocarbons.

*Food and Drug Statutes*

*The Food and Drugs Act.* The basic regulations for contaminants in food and drug products are found in regulations under the Food and Drugs Act. Use of certain substances as additives or in packaging is absolutely prohibited. Maximum quantity standards are prescribed for numerous toxic substances such as lead, arsenic and tin.

*Consumer Safety Statutes*

*The Hazardous Products Act.* This Act is a product-oriented statute which differs significantly from the Food and Drugs Act in that the basic Act does not provide broad powers or prohibition but rather relies upon the use of schedules to the Act and regulations to deal with specific hazardous substances. Schedule I to the Act prohibits the advertising, selling and importing of certain hazardous products such as furniture and toys created with a lead-containing material. Schedule II regulates the advertising, selling and importing of products such as cleaners containing chlorine compounds and glues containing ketone solvents.

*The Pest Control Products Act.* This Act is designed to control products used for the control of pests and the organic functions of plants and animals.

Detailed regulations have been promulgated under this Act prescribing registration requirements. The Department of National Health and Welfare provides advice respecting the toxicology of pesticides and related substances and is responsible for regulating residue levels of agricultural chemicals on food products under the Food and Drugs Act.

*The Transportation of Dangerous Goods Act.* This Act has been designed to promote public safety in the transportation of dangerous goods. The Act contains broad prohibition against handling, offering for transport or transport of dangerous goods unless all applicable safety standards are complied with and all containers, packaging and means of transport comply with all applicable prescribed safety standards.

*Statutes Regulating Specific Industries*

There are several groups of federal statutes that establish contaminant standards for particular industries. For example, the Pollution Prevention Regulations under the Canada Shipping Act: these regulations deal with discharge of contaminants by ships and are detailed enough to include, for example, a standard for lead discharge into coastal waters. Regulations under the Fisheries Act have established standards for chlor-alkali mercury plants, pulp and paper mills, and petroleum refineries.

*Federal-Provincial Interaction*

The complexity of the task of controlling toxic substances of past, present and future manufacture is such that cooperation by federal, provincial, and international agencies is essential. In order to make the most efficient use of resources, frank exchanges between all relevant agencies to eliminate redundant legislation, to develop needed legislation, and to analyze the quality of the inherent programs must take place.

The working mechanism we have used is the Federal-Provincial Advisory Committee on Environmental and Occupational Health. It advises the Ministers and Deputy Ministers of Health on all matters of environmental and occupational health. The Committee operates through a range of working groups on such topics as health surveillance of laboratory workers, indoor air quality, and drinking water.

The federal-provincial development of drinking water guidelines is a classic example of our cooperative approach. Traditionally, the control of drinking water quality in Canada has resided with the provinces and municipalities – although drinking water is defined as a food under the federal Food and Drugs Act.

For nearly 4 years, a federal-provincial working group – composed of scientists, technologists, and water engineers, – labored to produce the 1978 Guidelines for Canadian Drinking Water Quality. Limits for more than 50 parameters were given covering physical, chemical, microbiological, and radiological characteristics. The Guidelines include phenols (for organoleptic reasons) and 15 pesticides. Nitrilotriacetic acid was included because of its use (in excess of 25,000 t/annum in Canada) as a constituent of laundry detergents – most of which are disposed of in sewage. Trihalomethanes were included because of their production during disinfection. These guidelines have largely been implemented by the provinces and their commitment is surely a reflection of their own work in the genesis of the criteria and guidelines.

*Regulatory Reform*

The growth and pervasiveness of regulation in Canada and the prevailing political climate have prompted demands for regulatory reform. Many new regulations to protect the environment, occupational health, and consumer safety have been added in the last two decades. Today, despite some pressure for de-regulation, the momentum towards more regulation is very powerful, no less in Canada than elsewhere. It is noteworthy that the recent report by the Economic Council of Canada [1] chose to exempt the area of environmental pollution from their general conclusions to reduce and restrict regulation.

*Risk Assessment*

In any event, before controls are imposed, need and feasibility must be judged, preferably through the ordered sequence of risk identification, risk estimation, risk evaluation, and finally, risk management [5].

Initially, the hazard is recognized, then a scientific determination of risk is performed in as quantitative fashion as possible before the ultimately political evaluation is made on the acceptability of the risk and its consequences. This ordered approach helps separate the scientific estimation from the social evaluation and hence avoids some of the confusion that bedevils the risk assessment process.

Our national evaluations are, of course, buttressed by the international committees of World Health Organization (WHO) in cooperation with International Labour Organization (ILO), Food and Agriculture Organization (FAO), and United Nations Environment Programme (UNEP) and the process of risk estimation on an international scale should be strengthened by the new WHO/ILO/UNEP International Programme on Chemical Safety. This program, launched in 1980, should provide the international validity and cachet for the risk estimation of chemicals that is so urgently needed today.

How to choose which chemical to be tested from the multitude available and potentially toxic is a problem that has exercised many national agencies. In Canada, a List of Priority Chemicals under the Environmental Contaminants Act is gazetted annually by the Departments of Environment and National Health and Welfare. Three sets of criteria are applied to the selection of chemicals for the list: toxic effects to human health or the environment, persistence, quantity in use. Within these criteria, a committee by consensus develops a list. To date, a list of some 22 chemicals (and groups) has been established, classified as: Category I – regulations being developed; Category II – under investigation; Category III – further information required. In fact, this list is very similar to the top 50 chemicals list of the Interagency Testing Committee of the USA Toxic Substances Control Act.

*Examples of Risk Estimation in Canada*

The common trihalomethane after chlorination of drinking water is chloroform: a US National Cancer Institute rodent study found a dose-related incidence of malignant tumors in the kidneys of male rats and the liver of mice of both sexes. The daily dose to produce these effects was high – in the range 100–500 mg/kg body weight. In addition, studies with cough suppressants and mouthwashes have shown that hepatotoxicity occurs in humans at oral doses of chloroform between 1 and 25 mg/kg/day (70-kg person): occupational exposure to chloroform in the pharmaceutical industry is known to produce liver damage to the workers. *Tardiff* [4] carried out a detailed analysis of these data through four different models – margin of safety, and the statistical probit/log, linear one-hit, and two-step – to determine the maximum risk incurred by drinking tap water containing chloroform. A tenfold margin of safety applied to liver damage in man gives a maximum daily dose of 0.03 mg/kg. Extrapolation from the rodent studies gave the maximum risk at a maximum daily dose of 0.01 mg/kg range up to 0.4 cancers/million/year. For a 70-kg man consuming 2 liters of water daily, this intake represents a chloroform concentration of 0.35 mg/liter; this is the current recommended maximum acceptable concentration of total trihalomethanes in Canadian drinking water. Extrapolation from animal studies to man by means of statistical systems is defensible because we know the human target organ for chloroform and there is some evidence of similar metabolic pathways. On this basis, we can say that the risk of cancer to a member of the population from drinking water attributable to chloroform lies between no risk to kidneys and liver and a maximum of 1 in 2.5 million/year.

Lead is ubiquitous in our environment and has been the occasion for more studies and reviews than any other trace element. It has no known physiological functions and is widely distributed as a result of man's activities. By far the major source of airborne lead is the combustion of leaded gasoline: in Canada, some 70% of the emitted lead in 1972.

Examination of the Canadian data by *Hickman* et al. [2] of our integrated exposure to lead shows that the major contribution comes from food, although for children there is always the danger of ingestion of soil, dirt, or dust. These data were derived from analyses of total diets, drinking water from 70 municipalities, and from the National Air Pollution Surveillance. Under the Food and Drugs Act, tolerances for lead in food have been established, ranging from 0.08 ppm (infant formula) to 0.5 in wine. However,

Canadian intake is well below the WHO/FAO provisional tolerable intake for an adult of 430 μg/day.

The potential hazard of lead to vulnerable populations requires that emissions of lead into the air are progressively reduced. Under the Clean Air Act the maximum concentration of lead in gasoline was established at 3.5 g/gallon (0.77 g/l) in 1976.

There is a growing belief that the quality of indoor air may play a significant role in influencing health and that it may in fact be more important than the quality of the outdoor environment. We spend up to 70% of our time in nonoccupational environments and the impetus for energy conservation has led to a progressive sealing-up of our homes.

Increasing problems with urea formaldehyde foam insulation led to a ban on the sale of the product in December 1980. This decision was based on the report of an Expert Advisory Committee which noted the inherently unstable nature of the polymer leading to the release of formaldehyde. Formaldehyde is an irritant to the eyes, skin, and breathing passages and can produce allergic responses. Under controlled conditions, formaldehyde has been shown to induce nasal passage carcinoma in rats and mice. Mathematical models predict that exposure between 4 ppb and 1 ppm would produce 1 cancer in a million rats: the 1 ppm estimate is based on mathematical models that appear to fit the perceived mechanism of action of formaldehyde. The decision to ban the use of the urea formaldehyde foam was based not on a significant threat to life, but rather on the presumed risks of an inherently unstable product and the difficulty of developing controlling regulations.

The Department of National Health and Welfare has recommended an indoor air guideline of 0.1 ppm formaldehyde, similar to that adopted by some European countries: this level is one-twentieth of the workplace standard. If it is assumed that the cancer risk in man is similar to that observed in rats and mice, it can be calculated that the maximum additional cancer risk associated with an atmosphere containing about 0.05 ppm of formaldehyde – the background level in many homes – is equivalent to smoking one cigarette each week; the risk associated with 0.1 ppm formaldehyde is less than that of smoking one-third of a cigarette each day.

### *Societal Judgment*

These examples illustrate the manner in which a health authority uses risk estimation as a basis for regulatory control. The next state of risk assessment – risk evaluation – represents societal judgment and is ultimately political, beyond the realm of scientific analysis. Nevertheless before that judgment is made there are a number of serviceable and worthwhile techniques that can be used to provide a holistic approach whereby the total context of the problem is considered. A particularly valuable technique for risk evaluation is to compare the risk either with other risks, or the presumed benefit.

Comparing risks between different occupational groups, modes of transport, recreational activities, natural disasters and so forth in such terms as deaths/population unit/year or individual risks of cancer per year has been widely applied by *Pochin* [3] and *Wilson* [6]. When applied to disparate activities, devastating comparisons can be made. *Pochin* [3] gave a one-in-a-million risk of death to 400 miles travelled by air, three-quarters of a cigarette smoked, or 20 min of being a 60-year-old man.

Balancing risks from alternatives lies at the heart of the social evaluation process. It has been quantified for energy production. Qualitatively, the knowledge that replacement pesticides were available for DDT, replacement fluids for PCBs, other insulating materials for urea formaldehyde influenced the public health decisions to ban these chemicals.

Even when risks are well known and calculable there remains the problem of weighing them against the benefits giving rise to the risk. Different people place different values on things such as human life, esthetics, and national security. Many factors cannot be satisfactorily expressed in dollar terms because they involve important values on which there is no agreement. There is no question but that in the last decade we have made great progress in building a rational scheme on which to assess the risk from environmental chemicals. But, rationality is not enough. If it was, the public perception of hazards would equate with that of the professionals and the news media would have to find new sources of drama. The public is particularly exercised over risks over which they have no control and those which affect identifiable victims, particularly children. Inexorably, it is emotions at the deepest level that have been the level of innumerable legislative and administrative public health decisions.

Rational argument and logical thought are supreme in our ability to provide a structured, ordered system of risk assessment for chemicals. But they must not be taken too far: sometimes what passes for irrationality may be a well-founded suspicion of the operation of society. It is imperative that we recognize the limitations, as well as the powers, of the scientific method if we are to make sound judgments on the control of chemicals in our lives.

## *References*


1 Economic Council of Canada: Reforming regulation (Supply and Services, Canada 1981).

2 Hickman, J.R.; McBain, D.C.; Armstrong, V.C.: The contribution of drinking water to Canadian exposure to toxic substances. J. Envir. Monitor (in press).

3 Pochin, E.E.: Why be quantitative about radiation risk estimates?, Lecture No. 2, the Lauriston S. Taylor Lecture Series in Radiation Protection Measurements (National Council on Radiation Protection and Measurements, Washington 1978).

4 Tardiff, R.G.: Health effects of organics; risk and hazard assessment of ingested chloroform. J. Am. Water Works Ass. *69:* 658–661 (1977).

5 Whyte, A.V.; Burton, I.: Environmental risk assessment. Scope 15 (Wiley, Chichester 1980).

6 Wilson, R.: Testimony for OSHA hearings (1978).


E. Somers, Environmental Health Directorate, National Health and Welfare,
Tunney's Pasture, Ottawa, Ont. K1A OL2 (Canada)

Safety Evaluation and Regulation of Chemicals. 1st Int. Conf., Boston, Mass., 1982, pp. 26–30 (Karger, Basel 1983)

# Overview of Safety Regulations Governing Chemicals in Japan

*Yasuhiko Shirasu*
Institute of Environmental Toxicology, Tokyo, Japan

Since the tragedies of thalidomide and Minamata disease, safety regulations governing chemicals, including pharmaceuticals, pesticides, food additives, feed additives, and general chemical substances, have been tightened in Japan as elsewhere in the industrialized countries. In this paper the Japanese toxicology regulations, toxicological guidelines, good laboratory practices, and the present situation surrounding the Chemicals Control Law governing general chemical substances are surveyed.

## *Toxicology Regulations*

The main toxicology regulations are summarized in table I. Pharmaceuticals and veterinary drugs are controlled by the Pharmaceutical Law. Pharmaceuticals are dealt with by the Ministry of Health and Welfare (MHW) and veterinary drugs by the Ministry of Agriculture, Forestry and Fisheries (MAFF). Pesticides are controlled by the Agricultural Chemicals Control Law, primarily by MAFF with the collaboration of MHW and the Environment Agency (EA). Food additives are administered by MHW under the Food Sanitation Law. In 1975, a new Feed Safety Law was issued to control feed additives by MAFF. The Labor Safety Law controls chemical substances for occupational safety (Ministry of Labor, ML). In 1973, the Ministry of International Trade and Industry (MITI) issued the Chemicals Control Law to control chemical substances with the collaboration of MHW. Besides these laws, the Poisonous and Deleterious Substances Control Law regulates hazardous chemicals across-the-board.

*Table I.* Toxicology regulations

| Laws | Object | Ministry |
|---|---|---|
| Pharmaceuticals Law | pharmaceuticals | MHW |
| | veterinary drugs | MAFF |
| Agricultural Chemicals Control Law | pesticides | MAFF, MHW, EA |
| Food Sanitation Law | food additives | MHW |
| Feed Safety Law | feed additives | MAFF |
| Labor Safety Law | chemical substances | ML |
| Chemicals Control Law | chemical substances | MITI, MHW |

*Table II.* Toxicological testing guidelines

| | | | |
|---|---|---|---|
| 1 | Pesticides | 1972 | handling of registrations relating to the toxicity and residues of agricultural chemicals |
| 2 | Food additives | 1974 | genetic safety for food additives and others (proposed) |
| 3 | Pharmaceuticals | 1975 | reproductive effects (3-phase study) |
| 4 | Feed additives | 1977 | standards for evaluation of feed additives |
| | | 1980 | guidelines for studies on the basis of standards for evaluation of feed additives |
| 5 | Labor safety | 1979 | bacterial mutagenicity test for new chemical substances |

*Table III.* Pesticide toxicological requirement

| | |
|---|---|
| 1 | Acute toxicity studies |
| 2 | Chronic toxicity studies |
| 3 | Reproduction studies |
| 4 | Mutagenicity studies |
| 5 | Metabolism studies |
| 6 | Pharmacological studies |

### *Toxicological Testing Guidelines*

Major toxicological testing guidelines issued in the last 10 years are listed in table II. In 1972, MAFF issued guidelines for handling of registrations relating to the toxicity and residue of agricultural chemicals (table III). In these guidelines, requirements for acute studies, chronic studies, and reproductive studies are defined. No mutagenicity studies were required at the time, but they are now required by administrative guidance. Beside these

*Table IV.* Outline of standards for evaluation of feed additives

| | |
|---|---|
| 1 | Studies on effectiveness |
| 2 | Studies on residue |
| 3 | Acute toxicity studies |
| 4 | Short-term toxicity studies |
| 5 | Long-term toxicity studies |
| 6 | Succeeding generation studies |
| 7 | Teratogenicity studies |
| 8 | Oncogenicity studies |
| 9 | Genetic safety studies |
| 10 | Metabolism studies |
| 11 | Feeding studies using target livestock |
| 12 | Studies regarding development of resistant organisms |

studies, metabolism studies are needed. Some pharmacological studies are occasionally required.

In 1974, MHW proposed guidelines on genetic safety for food additives and others when controversy on mutagenicity and carcinogenicity of AF-2 (furylfuramide) was highlighted in the general public as well as in academic circles. These proposed guidelines cover various tests ranging from bacteria to mammals. These proposed guidelines influenced the mutagenicity tests of various guidelines which were issued afterwards. The toxicological testing guidelines for food additives were already issued by MHW in 1965.

In 1975, MHW issued guidelines for pharmaceuticals only on reproductive effects for 3-phase studies covering the following investigations: (1) fertility and general reproductive performance, (2) teratology, and (3) perinatal and postnatal study.

In 1977, MAFF issued 'Standards for Evaluation of Feed Additives' based on the new Feed Safety Law. The outline of the standards is summarized in table IV; 3 years later 'Guidelines for Studies on the Basis of Standards for Evaluation of Feed Additives' were issued. These guidelines are so far the most detailed toxicological guidelines provided by the Japanese Government. The reporting format is also included in these guidelines.

In 1979, ML issued bacterial mutagenicity test guidelines for occupational safety. They are only applied to new chemical substances and involve the use of *Salmonella typhimurium* TA98, 100, 1535, 1537, 1538, and *Escherichia coli* WP2 uvrA.

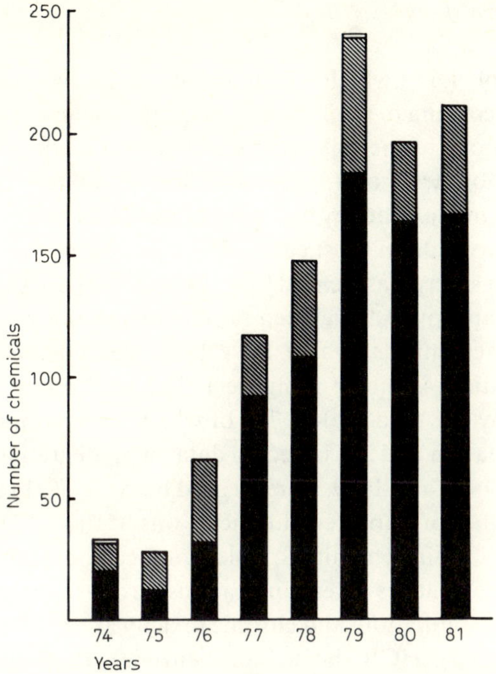


*Fig. 1.* Safety clearance of new chemicals from 1974 to 1981. The solid and shaded areas are the number of chemicals cleared by virtue of low bioaccumulation and good degradation, respectively.

### *Good Laboratory Practices (GLPs)*

Three kinds of GLPs are being drafted. The GLPs for pharmaceuticals have been drafted and will be issued soon by MHW. The GLPs for pesticides and general chemicals are in the drafting process by MAFF and MITI, respectively. The GLPs for pesticides are intended for toxicity studies. The GLPs for general chemicals cover not only toxicity studies but also studies for biodegradation and bioaccumulation. The GLPs have become a starting-point for international harmonization of regulations governing various chemicals. Such a trend is hoped to extend also to the various testing guidelines from the regulatory toxicology point of view.

*State of Chemicals Control Law*

The Chemicals Control Law which regulates common chemicals became effective in 1974. According to the Law, safety of new chemical substances should be cleared by one of the following test data: (1) biodegradation test, (2) bioaccumulation test, or (3) animal chronic toxicity test. However, clearance has been made either by biodegradation data or bioaccumulation data without toxicity data in most cases. The numbers of the new chemical substances of which safety was cleared from 1974 to 1981 are shown in figure 1. The numbers of approvals have been rapidly increasing year by year. The large number of chemicals, which have been cleared recently because of low bioaccumulation, is shown in figure 1. The total number of the new chemicals cleared by 1981 was 1,042, 779 of which were approved based on data on bioaccumulation and 261 based on data on biodegradation. Only two chemicals, each in 1974 and 1979, were cleared because of absence of chronic toxicity. The number of imported chemicals was 351 (33.7%) out of a total of 1,042. Among the existing chemicals, which totalled about 20,000, the seven specified chemical substances determined as unsafe due to their low biodegradation, higher bioaccumulation, and chronic toxicity are as follows: PCB (polychlorinated biphenyl), HCB (hexachlorobenzene), PCN (polychlorinated naphthalene), aldrin, dieldrin, endrin, and DDT.

Dr. Y. Shirasu, Institute of Environmental Toxicology, Suzuki-Cho 2-772,
Kodaira-shi, Tokyo 187 (Japan)

Safety Evaluation and Regulation of Chemicals. 1st Int. Conf., Boston, Mass., 1982, pp. 31–37 (Karger, Basel 1983)

# Chemicals Management in the OECD Context

*Joanne Kauffman Nichols*[1]

Environmental Consultant, Winchester, Mass., USA

The 24 Member countries of the Organization for Economic Cooperation and Development (OECD) account for two-thirds of world chemical production and exports. Most of this trade – about 90% – is intra-OECD. Any governmental action affecting chemical production and trade is bound to have repercussions in this highly competitive and interdependent industrial sector. The national economic implications of this are, of course, not insignificant. As an instrument of intergovernmental cooperation, the OECD is particularly geared to examine the economic effects of national actions. In this context, the chemicals issue is well suited to the Organization's mandate.

It is not surprising, then, that when national governments began to broaden the scope of chemical legislation in the early 1970s, the potential implications of the new generation of law went on the agenda of the Organization's existing Chemicals Group. Over the past 10 years, more than half of the OECD Member countries have initiated legislation aimed at the general control of chemicals. This means that governments increasingly will try to ensure that all chemicals are assessed for their potential risk to humans and the environment before they are put on the market.

The OECD Chemicals Program is designed to help Member countries cope with the significant demands this body of law places on government and industry. The overall aim is to develop policies that ensure the maximum protection from potential chemical hazards on man and the environment, while maintaining a viable chemical industry. The current OECD Chemicals Pro-

[1] *J.K.N.* is a consultant to the Chemicals Program of the OECD. The views expressed herein are those of the author and do not necessarily reflect those of the OECD or its Member countries.

gram focusses on three key areas for international harmonization: testing, evaluation, and information exchange.

Priority international issues in this field were identified at a meeting in 1978 on 'The International Control of Toxic Substances' sponsored by the Swedish Government [1]. The meeting was organized so that high-level officials from industrialized countries and international organizations could hold informal discussions on the international dimensions of the chemicals control issue. A clear articulation of interests and concerns ensued – and a framework for future international work was developed. The OECD was identified as the appropriate forum to carry out urgent work in six priority areas.

*OECD Chemicals Program*

Discussions at Stockholm focussed on four themes: assessment and evaluation of hazards (including risk assessments); potential risks for nontariff barriers to trade; information exchange needs including early warning systems; and the problem of confidentiality of data. Responding to the conclusion of that meeting, the OECD Council expanded its Chemicals Program in 1978 to consider specific issues within a 3-year time frame [2]. The major components of the program initially aimed at development of the following: (1) consistent data requirements and testing methods; (2) consistent standards for good laboratory practice and effective means of enforcing them; (3) mechanisms to facilitate information exchange, both with respect to substance-related data and to administrative actions; (4) solutions to problems associated with confidentiality of data, both to ensure international exchange of health and safety data, and to provide adequate protection for legitimate trade secrets; (5) consistent methodologies for analyzing the economic and trade impact of actions in regulating toxic chemicals; (6) a consistent international glossary of key terms.

The objective of these components has been to create an international system of testing and evaluation that will enable one government to accept the test data on a new chemical produced in another country.

A significant step toward achievement of that goal was the adoption in May 1981 by the OECD Council of the 'Principle of Mutual Acceptance of Data' [3]. The Principle asserts that data generated in the testing of chemicals in one OECD country in accordance with agreed test guidelines and quality controls would be accepted in all Member countries for the purposes of assessment.

The exercise of this Principle implies that whenever chemicals are tested, the test data should be scientifically reliable and of defined quality [4]. Until very recently, there has been little consensus on the information needed to evaluate the potential harm of chemicals, and even less agreement on the best means to develop that information. With the publication in 1981 of the OECD Test Guidelines and Principles of Good Laboratory Practice, this problem – in the international context – was mitigated [5]. These are tools which in practice provide a basis for producing scientifically reliable test data through the application of the current knowledge of science, and through good management of scientific work.

### *OECD Test Guidelines*

The OECD Chemicals Testing Program was launched in November 1977 as a response to the Council Recommendation of the same year instructing the Environment Committee to pursue a program of work on chemicals management. The program was given a 2-year mandate to consider the state-of-the-art of chemicals testing – and to develop guidelines for testing in four major areas: physical and chemical properties; degradation and accumulation; health effects, and ecotoxicology. Lead countries offered to host Expert Groups to consider test methods in each area. Two groups were formed to consider health effects testing: one on short-term toxicology and another on long-term toxicology. A sixth group was formed to consider the entire concept of step-sequence testing, including the basic data package considered necessary to make an initial assessment, as well as criteria for moving on to more complex testing.

Between 1977 and 1980 more than 200 nationally nominated experts from industry, government and academia met in capitals of the Lead Countries for each expert group to develop the most current, internationally preferred testing methods available.

The selection methodology involved the identification of priority issues in each of the testing areas, and an exchange of national experiences among experts who reviewed the scientific documentation on available methods and proposals. Frequently, laboratory intercomparison testing programs, or ring tests, were carried out to determine the practicability of the proposed methods. Laboratories from several OECD countries ran the tests and compared results. On the basis of their reports weaknesses could be identified and improvements recommended.

These were then subject to an extensive peer review process throughout the OECD membership. National authorities were asked to seek appropriate technical review and comment on each of the testing areas. A Review Panel was then set up under the aegis of the Organization's Chemicals Group to review the comments and amend the Guidelines accordingly. The 51 completed guidelines were selected on the basis of technical and economic practicability. They are aimed at revealing the physical characteristics of chemicals, their chemical reactivities, their ability to degrade or accumulate in man and the environment, and their interactions with species in various ecosystems. Approved by the OECD Council in 1981, they are now in use in Member countries.

The Guidelines were not, however, the only result of this program. In particular, the Expert Groups contributed significantly to the knowledge base on chemical effects testing and to the identification of gaps and weaknesses in testing areas. Most importantly, this work is a first step in the achievement of scientific consensus on the methods and means to assess chemicals.

In light of the dynamic nature of the art of chemical testing, an Updating Program was also established in 1981 to keep the OECD Test Guidelines current with changes in the state-of-the-art and with advances made in the relevant sciences. New guidelines will be developed in areas not yet covered, and proposals will be considered for new or modified tests which offer conspicuous advantages over those already adopted. Some of the areas being considered in the Updating Program are genetic and reproductive toxicity, neurotoxicity, ecotoxicological tests with other species such as bacteria, birds and earthworms, and abiotic degradation of chemical substances.

It should be noted that these Test Guidelines are not prescriptive or rigid test protocols. One guideline may, for example, involve several alternative methods. It is axiomatic that scientists examining chemicals for potential effects need room to exercise expert judgment.

The publication of the OECD Test Guidelines does not mean that the international challenges in testing have all been met. The Updating Program is, in fact, recognition that in some critical areas tests have yet to be developed or internationally accepted. Moreover, there remains the challenge of developing more cost-effective in vitro tests which provide reliable guides to real-life situations.

The Test Guidelines provide guidance on the nature of tests – the first characteristic, if you will, of valid test data. The second is good management of scientific work. The issues surrounding quality assurance in testing sound simple, but the scientific and social judgments are complex.

*Good Laboratory Practices*

The OECD Principles of Good Laboratory Practice were developed in a similar Expert Group process to that used in development of the Test Guidelines. And the OECD Principles of Good Laboratory Practice were also adopted by the Organization's Council in 1981.

GLP is seen as a management tool. The aim is to demonstrate the organization of a test facility with the objective of maintaining and promoting the quality of test data. The Principles deal specifically with the organization, practice, and technical conditions under which laboratory studies are planned, performed, monitored, recorded and reported. It promotes practices and procedures which sustain the careful planning of a test and its accurate and complete documentation.

The Principles extend to all categories of testing. They have been developed for the testing of industrial chemicals – which are the focus of the OECD's Chemicals Program – but they can be applied to the testing of chemicals which previously have been regulated such as pharmaceuticals, pesticides, food additives, etc. If followed, the Principles will assure that high quality data are produced consistently in the testing of chemicals. The next step is to build a system which allows mutual recognition by nations of one another's schemes for compliance with these principles. This is, in fact, a current challenge of the Expert Group.

*Hazard Assessment*

Not every chemical entering the market must or should pass through a complex and comprehensive group of tests. A much debated question in the policy arena of the general control of chemicals is, in fact, 'How much data should be generated on each new chemical in order to make an assessment of the risks it might pose if introduced onto the market?' This question has been considered by the OECD Expert Group on Step-Sequence Testing.

The Group developed what they call the Minimum Pre-Marketing set of Data – or MPD as it is abbreviated – to be used in making a first assessment of a chemical. A very similar approach has already been adopted by the European Communities through its 1979 Directive on premarketing notification of chemicals. In developing the MPD, the Expert Group determined which of the tests considered in the Testing Program could be used to produce data necessary to make a first crude assessment of any chemical before it enters the market, while, of course, allowing for flexibility in making individual decisions.

The value of the MPD lies in its ability to indicate whether a new chemical is likely to be hazardous. This constitutes, in effect, the first step in the scientific hazard assessment process. If initial data on a given chemical indicate that the chemical is potentially hazardous, then the decision must be made whether to regulate immediately, or to require further testing – and, if the latter case, what testing? To answer this question in a systematic way, the Expert Group that produced the MPD is considering the concept of step-sequence testing to determine criteria for moving up the scale to increasingly complex testing.

The OECD group defines hazard as a 'function of two broad considerations: the potential of the chemical to harm biological systems, and its potential for exposure'. The ability to evaluate and to relate the various data components relevant to these considerations is thus essential to a determination of hazard. This is a complex process – and there is growing recognition in both government and industrial forums of the need to establish a rational and consistent framework for hazard assessment. One aim of current OECD work in this area is the development of principles for the evaluation of data elements in MPD that may be used as a guide in the assessment of chemicals.

Three working parties (again using the Lead Country approach) have been established to consider principles for assessment of individual data elements in each of the following areas: exposure analysis; natural environment effects, and health effects. The working parties are also considering combinations of the various data elements and approaches to integrated evaluations.

The development of analytical approaches to hazard assessment is in an early stage. Currently, there is much activity in both national and international bodies to develop principles to guide hazard assessment. There is the possibility that mutual acceptance of hazard assessments may eventually be achieved in much the same way that earlier work on testing led to adoption of the Principle of Mutual Acceptance of Data. It is clear, however, that principles for hazard assessment (once established) would need to be frequently reviewed and updated in the light of practical experience.

### *Future Directions*

The scientific aspects of chemicals management which have been and are being explored within the OECD Chemicals Program are basic elements in assessing chemicals. The outcome of these assessments will determine the basis, to a large extent, of setting conditions on the use, marketing, storage, handling and disposal of these substances.

The work to date has been aimed at new chemicals management. But, in fact, the principles are relevant to the problems of existing chemicals as well. Recently, the mandate of the Chemicals Program was extended through 1984 to consider further hazard assessment, the many international issues related to the control of existing chemicals, and elements of decision-making.

A benefit of the OECD harmonization efforts has been to help minimize the uncertainty that has characterized decision-making on chemicals. This uncertainty has been accentuated in the past by a lack of scientific consensus on the means to evaluate and assess chemicals, and by the absence of frameworks for analyzing and evaluating the various elements – both scientific and economic factors – to be considered in decision-making.

Progress has been made to remedy this uncertainty, but we are still in the early stages of finding answers to the many questions inherent in the control of chemicals. In seeking such answers jointly – and as openly as possible – nations can hope to develop efficient policies that take account of the many factors needed to determine and weigh chemical risks against economic and environmental benefits.

## *References*


1 'Control of Toxic Substances': Proceedings from an International Meeting arranged by the Swedish National Products Control Board, Hasselby Castle, Stockholm, April 11–13, 1978.

2 Decision of the Council 'Concerning a Special Programme on the Control of Chemicals'; adopted September 21, 1978 (OECD, Paris 1978).

3 Decision of the Council 'Concerning the Mutual Acceptance of Data in the Assessment of Chemicals'; adopted May 12, 1981 (OECD, Paris 1981).

4 For further discussion of harmonization, see Crawford, P.J.: The OECD approach to chemical control. Toxic Substances J. *3:* 62–70 (1981).

5 OECD Guidelines for Testing of Chemicals (Organization for Economic Cooperation and Development, Paris 1981).


Joanne Kauffman Nichols, Environmental Consultant, 30 Hillcrest Parkway, Winchester, MA 01890 (USA)

Safety Evaluation and Regulation of Chemicals. 1st Int. Conf., Boston, Mass., 1982, pp. 38–43 (Karger, Basel 1983)

# WHO's International Programme on Chemical Safety: Evaluation of the Effects of Chemicals on Health

*Laila A. Moustafa*

WHO International Programme on Chemical Safety, Interregional Research Unit, Research Triangle Park, N.C., USA

Concern for health-related environmental problems is likely to increase both in the developed and developing countries due to the associated demand for more food, consumer products, energy, transportation, etc. There is not only the problem of increased waste production but also that of rapidly accelerating utilization of chemicals in all fields. All in all it is estimated that today at least 60,000 chemicals are used in practical applications. Together with unwanted waste or by-products from man's activities, these chemicals find their way into the air, water, soil and food in the course of their production, transport, use or disposal [4].

Environmental regulations have led some industries to site new installations in places where environmental policies are nonexistent or less severe [2]. However, it is not for scientists to answer the question, 'How much risk is acceptable with respect to a given chemical hazard?' Socioeconomic and other factors may very well have an even more important role to play and these often differ considerably from country to country. Some developing countries have been tempted to accept 'polluting industries' in order to expand their economies. Every society has to do its own decision-making using its own political instruments – once the scientific community has given its advice on a potential or real chemical hazard.

The United Nations Environment Program (UNEP) launched a series of executive summaries on environmental themes in order to provide a convenient reference for the educators, scientists and decision-makers of the world. The first in this series, *In Defense of the Earth* [3], contained the three basic documents upon which the fundamental principles of environmental management are based: The 1971 Founex Report, the 1972 Stockholm Plan of Action and the 1974 Cocoyoc Declaration.

### *WHO Programs in Environmental Health*

Health problems arising from exposure to pollutants and other hazards in the environment have long been among the concerns of the World Health Organization (WHO). However, it was not until the early 1970s that these aspects gained increased prominence within the organization's program on promotion of environmental health. This rise in emphasis coincided with the 1972 Stockholm Conference on the Human Environment. The conference's recommendations urged the national authorities and international bodies to give increased attention to environmental problems.

The World Health Assembly (WHA) responded with resolutions in 1973 (WHA Resolution 26.58) and 1974 (WHA Resolutions 27.49 and 27.50) which provided guidance and support for the development of the long-term WHO program in environmental health with emphasis on (1) the assessment of the effects of environmental conditions on health; (2) development of systems for monitoring the pollutants and other environmental factors that may be harmful to health in air, water, food and the working environment; (3) assistance to member states in planning and implementing environmental control programs; (4) preparation of criteria documents on the effects of biological, chemical and physical agents on man, and (5) promotion and coordination of research on the health effects of environmental pollutants and other environmental factors.

#### Environmental Health Criteria Program (EHC)

The EHC was started in 1973 to provide evaluations of health risks associated with exposure to selected environmental agents. The assessment of health effects of environmental agents and the development of environmental health monitoring systems were identified for priority action. Consequently, WHO, with its own resources and supported by the United Nations Environment Fund, began to plan and initiate new projects in these two areas of concern (table I).

On the monitoring side, a global project was initiated in 1973 on air quality monitoring followed by projects on water and food contamination monitoring. A few years later, pilot studies were started on biological monitoring and on detailed assessment of exposure to air pollutants. These last two pro-

*Table I.* Existing 5 global projects on environmental monitoring (developed and implemented by WHO/UNEP in cooperation with other international organizations)

1. Air Pollution Monitoring (in cooperation with WMO)
2. Global Water Quality Monitoring (in cooperation with WMO and UNESCO)
3. Joint FAO/WHO Food and Animal Feed Contamination Monitoring Programme
4. Pilot Project on Assessment of Human Exposure to Air Pollutants (in cooperation with WMO)
5. Pilot Project on Assessment of Human Exposure to Pollutants through biological monitoring

jects provide the initial steps toward human exposure assessment and are gradually being developed as the health-related component of the Global Environmental Monitoring System (GEMS). Although the three projects on air, water and food monitoring were somewhat differently planned and structured, they all had essentially the same objectives, namely: (1) technical cooperation with member states in their efforts to establish and develop environmental monitoring systems to protect human health from adverse environmental influences, and (2) the international exchange and interpretation of information on levels and trends of environmental pollution and other hazards.

This information is intended to define which problems are widespread and would benefit from international cooperation in their health risk evaluation or control strategy development. It should also increase both public and governmental awareness of the problems at hand. That this information can contribute to the evaluation of regional problems such as long-range transport of air pollution or pollution of international rivers should also not be overlooked. Although the results of this work are for the most part encouraging, global networks of this type must be supported by a continuing and comprehensive quality assurance program. The monitoring program is a collaborative activity involving other international organizations such as the World Meteorological Organization (WMO), the United Nations Education, Science and Cultural Organization (UNESCO) and the Food and Agricultural Organization of the United Nations (FAO) [4]. The different networks were designed with certain general considerations and constraints in mind as follows: (1) Type and extent of environmental problems, by country and by region; (2) achievement of global coverage; (3) financial and personnel resources available for implementation of the projects, and (4) interest and existing capability for participation in the program.

International Programme on Chemical Safety (IPCS)

The 31st WHA (1978), after reviewing the Director-General's report on WHO's human health and environment program (document A31/27), adopted resolution WHA 31.28 on the evaluation of the effects of chemicals on health (WHO Official Records, No. 247, 1978, p. 18). In that resolution, the WHA requested the Director-General of WHO to further promote international cooperation in evaluation of the health effects of toxic and hazardous chemicals. As a step in implementing resolution WHA 30.47, the IPCS was established in 1980 in cooperation with UNEP and the International Labour Organization (ILO) to accelerate and expand the preparation of criteria documents and related activities. The program was based on ongoing national and WHO activities and included relevant activities of the International Agency for Research on Cancer (IARC) and possibly other international work related to chemical safety [1]. The main outputs of the IPCS:

(1) Evaluations of the effects of chemicals that are relevant to human health (including special effects such as carcinogenesis, mutagenesis, and teratogenesis, and acute and chronic effects on different organs and organ systems). Such reports might vary in form from comprehensive criteria documents to single sheet reports and risk assessments of new chemicals. It should be recognized that environmental factors of physical and biological origin may interact with chemicals and in some cases enhance their effects. Such interaction should be borne in mind when health risks of chemicals are being evaluated.

(2) Guidelines on exposure limits, such as acceptable daily intakes for food additives and pesticide residues, and tolerances for toxic substances in food, air, water, soil and the working environment.

(3) Guidelines on appropriate methods for exposure measurement and assessment, toxicity testing, epidemiological and clinical studies, and risk assessment and hazard evaluation.

(4) Information on methods for coping with chemical accidents; case studies of specific past accidents and emergencies; and intervention in emergencies, on request. The dissemination of information being a priority of the IPCS, it was agreed that the Central Unit would develop, in close cooperation with ILO and the UNEP International Register of Potentially Toxic Chemicals (IRPTC) [2], appropriate procedures to that end within the IPCS.

(5) Advisory services for technical cooperation, and information on regulatory and other control chemicals, and on other subjects relevant to technical cooperation in this area.

(6) Trained manpower.

The WHO Regional Office for Europe (EURO) is involved in a regional project on planning for emergencies and accidents involving the release of toxic chemicals. EURO functions as an extension of the Central Unit in this project which, as with everything else in the IPCS, would be implemented through the Lead Institutes (LI) whenever possible. The objective is to enable countries without appropriate plans to deal with such situations [5–7]. The manpower development program includes: (1) definition of the toxicologist in terms of the tasks he/she performs and the skills needed to perform them; (2) a survey of the needs for toxicologists; (3) a survey of existing training programs, and (4) the design of relevant curricula.

Measurement Methods and Quality Assurance

Any monitoring scheme which involves many laboratories is faced with questions concerning the comparability and validity of the data which are collected. Often laboratories employ somewhat different measurement methodologies and have different degrees of quality assurance. This problem is compounded in an international setting, where even greater differences in monitoring approaches, methodologies and quality assurance exist among the laboratories located in different countries.

Through international cooperation, harmonization of measurement methodologies and validity of data can be improved through interlaboratory quality assurance studies. Furthermore, for developing countries in particular, participation in a global network provides possibilities for information exchange with programs in other countries.

WHO and the International Council of Scientific Unions/Scientific Committee on Problems of the Environment (ICSU/SCOPE) convened a meeting in June 1978 in Bellagio, Italy, to consider means of promoting the use of appropriate methodology for the evaluation of the safety of chemicals and to improve the scientific basis for such evaluations. It recommended that a scientific group on methodologies for safety evaluation of chemicals be established under the sponsorship of WHO, ICSU/SCOPE and other interested organizations. Its functions would be (1) the study, review, evaluation and publication of reports on selected topics and (2) provision of advice to sponsoring organizations. The group was established initially for 3 years and was open to co-sponsorship by other organizations.

The IPCS-proposed program on methodology complemented those of the Organization for Economic Cooperation and Development (OECD) on test methodologies. It goes without saying that the international organizations engaged in elaborating test methodologies are in need of close cooperation to avoid unnecessary overlap.

In closing, it is our sincere hope that WHO can be of use to all nations, for genuine partnership between member states and WHO is the key to attaining international health goals. This common goal can be achieved by making use of WHO to the maximum, using it to facilitate such cooperation among nations, in order to reach practical agreements.

*References*

1 International Programme on Chemical Safety (IPCS): Report of the 2nd session of the technical committee. EHE/81-21 (WHO, Geneva 1981).

2 United Nations Environment Programme: International register of potentially toxic chemicals. IRPTC Bull. 4: No. 2 (1981).

3 UNEP: In defense of the earth: the basic texts on environment – Founex, Stockholm Cocoyon. Executive Series 1 (United Nations Environment Programme, Nairobi 1981).

4 WHO: Consultation on health-related environmental monitoring. EFP/81-12 (WHO, Geneva 1981).

5 WHO/EURO: Manpower development for control of chemicals, occupational profiles and curriculum design. Interim Document 2 (WHO, Regional Office for Europe, Copenhagen 1981).

6 WHO/EURO: Planning emergency response systems for chemical accidents: administrative guidelines. Interim Document 1 (WHO, Regional Office for Europe, Copenhagen 1981).

7 WHO/EURO: Progress report, July 1981. Interim Report 3 (WHO, Regional Office for Europe, Copenhagen 1981).

Dr. Laila A. Moustafa, WHO International Programme on Chemical Safety,
Interregional Research Unit, PO Box 12233, NIEHS Mail Drop A2-06,
Research Triangle Park, NC 27709 (USA)

Safety Evaluation and Regulation of Chemicals. 1st Int. Conf., Boston, Mass., 1982, pp. 44–49 (Karger, Basel 1983)

# A Common European Approach to Toxicological Assessment: The Work of the Council of Europe

*H. Scicluna*

Council of Europe, Strasbourg, France

Established by ten nations on May 5, 1949, as Europe's first political institution, with the first international parliamentary forum in history, the Council of Europe has today 21 member states: all the countries of Western Europe, except Finland. The aim of the Council of Europe is to achieve a greater unity between its members for the purpose of safeguarding and realizing the ideals and principles which are their common heritage, and facilitating their economic and social progress. The pursuit of that objective is the responsibility of two bodies – the Committee of Ministers and the Consultative Assembly.

The scope of the Council of Europe's competence under its statute is vast, since only defence questions are excluded. Its activities touch on every aspect of life in Europe and find expression in many forms, such as recommendations and resolutions addressed to, and conventions and agreements concluded between, all member states. Where, however, a lesser number of states wish to engage in some action in which not all their European partners desire to join, they can conclude a 'partial agreement' which is binding on themselves alone.

The work in the field of toxicology is carried out within the framework of a partial agreement between Austria, Belgium, Denmark, France, the Federal Republic of Germany, Ireland, Italy, Luxembourg, the Netherlands, Norway, Sweden, Switzerland and the United Kingdom. It relates exclusively to food (flavorings, plastic packaging materials, pesticides) and cosmetics.

*Regional Cooperation*

The need for a common approach to toxicological assessment is the result of various factors, among which are the increase in the volume of trade, rapid communication and the consumers' sensitive awareness of the necessity for health protection. Harmonization has, however, proved difficult and progress has been slow, mainly due to two reasons: (1) an exaggerated empirical approach in the attempt to harmonize standards without previous agreement on the basic principles underlying these standards, and (2) a worldwide approach to the problem, without sufficient consideration of the socioeconomic differences between the various countries.

The work carried out and the results achieved in the public health field by this restricted number of members of the Council of Europe clearly illustrate the effectiveness of regional cooperation on a scientific and flexible basis. These countries have comparable levels of industralization and social standards. They are, therefore, confronted with the same problems, and due to a common socioeconomic and cultural background, they tend to adopt similar basic principles in order to resolve them.

*Objective*

The main objective of the Council of Europe in its work on toxicological evaluation is exclusively the protection of public health. Reference standards are therefore elaborated on a purely scientific basis without undue consideration of economic and other factors, and on the basis of criteria and following guidelines which provide the best protection against health hazards.

*The Positive List Approach*

The elaboration of reference standards in flavorings, plastic food packaging materials and cosmetics is based on the positive list approach. According to this system, only those substances which can be considered as safe from the public health point of view are listed. The third edition of the Council of Europe's publication *Flavouring Substances and Natural Sources of Flavourings* [1] states that an effective legal control must be based on positive lists. This philosophy has been equally applied in all the other fields of toxicological evaluation as the best suited for the protection of public health. The

Commission of the European Communities has adopted the same approach in the preparation of its directive in this field.

### *Practical Application*

If the positive list system provides a rigorous scientific approach, it is no less important to ensure a practical application of the toxicological evaluation made. This requires a flexible approach.

Toxicological evaluation, particularly if the positive list approach is strictly adhered to, involves various difficulties. In the first place, there is the large number of compounds which have to be assessed within a reasonable time. With a complex situation in both the technological and toxicological field, experts have to rely on a variety of data (levels of technological use, chemical structure and analogy to compounds of similar or related constitution and knowledge of biochemical behavior, actual toxicological information) to arrive at a compromise. Safety margins have therefore to be established within reasonable limits and it is often necessary to rely on 'common sense' arguments such as the harmlessness of a product as evidenced by prolonged use. In the third edition of the above-mentioned publication [1] a number of natural sources of flavorings are listed as temporarily acceptable, in view of their long history of use without evidence of acute untoward effects.

The same realistic approach is adopted in classifying as permissible a number of flavoring substances for which there is a paucity of available technological and toxicological data. It was felt that it would be unrealistic at the present moment not to admit the temporary continued use of some of these compounds if a reasonable assumption as to their safety could be made, pending the receipt of further relevant information. The same policy of classification is followed in the toxicological evaluation of those cosmetic ingredients where more information is needed before assessment of health hazards can be completed, but for which there is sufficient evidence to indicate that there is no hazard.

### *Guidelines for Toxicological Assessment*

Within the framework of a flexible positive list system, a number of practical guidelines are used as a basis for toxicological assessment of the potential health hazards of the compounds.

In all fields – flavoring substances, plastic packaging materials, pesticides, cosmetics – the assessment of the safety in use is based on the results of studies in animals and observations in man. These include studies by other intergovernmental organizations such as the World Health Organization (WHO) and the Food and Agriculture Organization (FAO), as well as non-governmental organizations and national laboratories in Europe, USA, Japan and other countries. A practical parameter equally applicable to all is the comparability of the chemical structure with that of compounds of known toxicological and biochemical properties. With regard to food toxicology, the acceptable daily intake for man recommended by the Joint FAO/WHO Expert Committee on Food Additives has been adopted. Other parameters taken jointly with the above factors vary according to the nature and purpose of the substances considered.

We can distinguish between three types of parameters applicable, respectively, to intentional additives which are ingested (flavorings), contaminating compounds (plastic packaging materials, pesticides) and substances which are intended for external use only (cosmetic products).

In flavorings [1], which are intentional additives and which are ingested, factors considered are:

– available information on levels of use in various dietary components;

– the natural occurrence in staple items of the normal diet;

– the preferential consumption by specially vulnerable groups of consumers;

– the contribution of a flavoring to the total dietary load from all sources;

– the presence of self-limiting characteristics, such as organoleptic properties.

With regard to products which may leave harmful residues, we can distinguish between two categories of substances: plastic packaging materials and pesticides. In both cases, the main parameter is the possibility of residues, and eventually the tolerable level of those residues.

In the case of substances used in plastic packaging materials [2], they have been classified on the assumption that they migrate in unchanged form into food. The migration and toxicological significance of compounds formed from the originally added substances during the production, processing or use of plastic materials has not been considered.

In the field of pesticides [3], the Council of Europe has not carried out any toxicological evaluation of pesticide residues, but has limited itself to giving detailed general guidelines for use by manufacturers concerned with the registration of agricultural and nonagricultural pesticides. The main distinguishing feature of pesticides is that they are used in an open environment. Consequently, a parameter particular to their toxicological evaluation is the precaution to be taken to prevent harm to domestic animals and wild life, including bees and fish.

Cosmetic products [4] constitute a third category of chemical substances studied by the Council of Europe. Cosmetic products are by definition intended only for external use. Cosmetic ingredients should not demonstrate such a degree of pharmacological activity that the cosmetics in which they occur would be classed as pharmaceutical preparations. The parameters therefore basically concern irritancy and sensitization, but as ingredients and final products vary widely in their nature and mode of use, it is not possible to have a uniform system of evaluation. The intended mode and frequency of use must be kept in mind, as well as other foreseeable areas in which the final product may be used under normal conditions of use. Therefore, while it is always necessary to look for evidence of irritation and allergic sensitization, it will also be necessary in some instances to examine for evidence of systemic toxicity where an ingredient from the final product may be absorbed through the skin or mucous membranes, inhaled or ingested.

It should be added that the toxicological potentials of all these chemical substances have been evaluated without consideration of possible toxic interaction among themselves, or between them and other constituents of food, other chemicals in the environment or substances used for medical purposes. The Council of Europe experts have, however, drawn the attention of the authorities to the possible hazards resulting therefrom.

*Legal Application*

The intergovernmental work of the Council of Europe is put into effect by:

– a convention or an agreement signed by the contracting states, which, as a general rule, prior to its entry into force, is ratified by governments after having been referred to the parliamentary organs of the signatory states;

– a resolution of the Committee of Ministers which is transmitted to the member states for implementation;

– the simple publication of a report which sets out the views of the expert committee on a particular subject.

The results of the toxicological evaluation carried out by the Council of

Europe normally take the form of a scientific report. These reports are not formally adopted by the Committee of Ministers. They constitute, however, the common viewpoint of the best governmental experts from the member states and as such are scientifically valid and authoritative.

Food standards are of a highly technical nature and largely dependent on a number of factors which vary from country to country. They can therefore be successfully harmonized on an international level if there is a sufficiently flexible procedure allowing for this. To facilitate this process of harmonization, the Council of Europe has drawn up reference standards which each participating state can adapt to the extent and in the manner best suited to meet the conditions prevailing in its country, and which can easily be updated in accordance with scientific development. For this purpose, the Council of Europe has avoided the use of rigid legal instruments which create binding obligations and has chosen flexible methods of application.

### *References*

1 Council of Europe: Flavouring substances and natural sources of flavourings; 3rd ed. (Maisonneuve, Moulins-les-Metz 1981).
2 Council of Europe: Substances used in plastics materials coming into contact with food; 2nd ed. (Council of Europe, Strasbourg 1982).
3 Council of Europe: Pesticides; 5th ed. (Council of Europe, Strasbourg 1981).
4 Council of Europe: Cosmetic products and their ingredients; lst ed. (Council of Europe, Strasbourg 1978); addendum (1981).

H. Scicluna, Council of Europe, BP 431 R6, F-67006 Strasbourg Cédex (France)

Safety Evaluation and Regulation of Chemicals. 1st Int. Conf., Boston, Mass., 1982, pp. 50–57 (Karger, Basel 1983)

# Regulating Environmental Pollutants: A View from a State Agency

*Halina Szejnwald Brown*

Office of Criteria and Standards, Massachusetts Department of Environmental Quality Engineering, Boston, Mass., USA

Since the landmark year 1970 when the Clean Air Act was adopted by the United States Congress, several other fundamental pieces of legislation have been adopted. These were: Safe Drinking Water Act of 1974; Clean Water Act of 1974; Resource Conservation and Recovery Act of 1976 (RCRA) and Toxic Substance Control Act of 1977 (TSCA).

The Massachusetts Department of Environmental Quality Engineering (DEQE), established in 1975, is the Commonwealth's environmental regulatory agency. Many of the activities which relate to air and water quality and hazardous wastes are required from DEQE by one or more of the above legislative acts.

Most of the federal primary standards for air and water quality have been adopted by the Massachusetts DEQE unchanged. These standards are presently used in regulatory and enforcement decisions such as permit issuance, water supply operations, etc. However, EPA primary air and water quality standards cover only a small fraction of chemicals which can enter the environment. Presently, there are 17 primary drinking water standards and 6 primary ambient air quality standards. There are no federal regulations which would put a ceiling on the potential environmental human exposure to most commercial chemicals.

There are two important time factors involved in establishing an environmental quality standard. First, a pollutant is chosen for regulation. Secondly, a complex and time-consuming regulatory process takes place. The standard-setting machinery is a slow moving one. For instance, action which led to promulgation of the drinking water standard for trihalomethanes in November 1979 started in late 1974 with the discovery of these chemicals in drinking water. The proceedings leading to promulgation of the National Ambient Air Quality Standard for lead in October 1978 started in 1976 with a court order to list lead as a criteria pollutant. It is estimated that the average time involved in rulemaking by EPA is now approaching 4 years. This is not surprising considering that, in order to arrive at the final rulemaking for an average pollutant, EPA goes through extensive review of scientific literature, publication of the advance notice of a proposed rulemaking, issuance of the proposed standard, usually more than one public comment period, assessment of the economic and social impact of the proposed standard and technical details of implemention of the standard.

Despite its length, the regulatory process, once started, usually culminates in promulgation of a standard. There are, however, tremendous delays in the initial triggering of the regulatory activity. Officials at the federal agencies are often reluctant to take the first step to regulate a toxic substance because they know what happens next: outside pressures are applied, both directly and through Congress, followed frequently by a lawsuit, and shortages in resources become acutely obvious. The most experienced regulators are intimidated by the complexity of the task. As a result, regulatory agencies typically do not act until forced to do so. Finally, with the immense variety of potentially harmful chemicals in the environment, the chances for a given agent to be chosen for regulation are relatively small.

*Trihalomethanes*

In 1974, it was reported that the presence of humic and fulvous substances in waters treated with chlorine for disinfection purposes lead to formation of chloroform and other trihalomethanes. In late 1975, acting in compliance with the Safe Drinking Water Act, EPA initiated the National Organic Monitoring Survey (NOMS) in 113 cities. The NOMS demonstrated that considerable amounts of trihalomethanes could form in treated water after it entered the distribution system on the way to the consumer's tap. It also showed that trihalomethanes far exceeded the concentrations of other synthetic organic contaminants in finished drinking water (chloroform ranging from 0.001 to 0.540 mg/l). In July 1976, EPA published an advance notice of proposed rulemaking which ultimately led to promulgation of the present standard of 0.100 mg/l for trihalomethanes.

*Lead*

Lead is emitted into the atmosphere by vehicles burning leaded fuel and by certain stationary sources. EPA's initial approach to controlling lead in the air was to limit the lead emissions from automobiles, its principal source. In 1975, the Natural Resources Defense Council (NRDC) and others brought suit against EPA to list lead as a pollutant for which air quality criteria would be developed and NAAQS established. The Court ruled in favor of NRDC. EPA listed lead on March 31, 1976, and the present standard of 1.5 $\mu g/m^3$ was promulgated in October of 1978.

*129 Priority Pollutants*

A June 1976 court settlement involving EPA versus several environmental groups (NRDC and others) has commonly become known as the 'EPA Consent Decree'. These groups brought suit against the Agency for failing to implement portions of the Clean Water Act. One of the outcomes of the litigation was the requirement that EPA publish a list of 65 toxic compounds or classes of compounds for which technology-based effluent limitations and guidelines would be required. The original list compiled by EPA was nonspecific and practically impossible to comply with on the analytical level. In the process of resolving this and other problems, a larger and much more detailed list of 129 compounds and groups of compounds was produced in June of 1978, now known as the priority pollutants.

*Ambient Water Quality Criteria*

The Clean Water Act requires EPA to publish, and periodically update, water quality criteria. These criteria are to reflect the latest scientific knowledge on the identifiable effects of pollutants on public health and welfare, aquatic life and recreation. The Act was amended in 1977 following the 1976 EPA Consent Decree. In the amendments, Congress designated the 65 pollutants listed in the Consent Decree as subjects of such criteria. Consequently, the advanced notice of proposed rulemaking was published by EPA, followed by the 65 water quality criteria.

*Suggested No Adverse Response Levels (SNARL)*

Suggested No Adverse Response Levels are guidelines for drinking water quality issued by the EPA Office of Drinking Water upon request. They are not legally enforceable standards and they may or may not lead ultimately to the issuance of a national drinking water standard. SNARL is a level of a contaminant in drinking water at which adverse health effects would not be anticipated, with a margin of safety. These guidelines are frequently issued in response to emergency requests from states. Presently, there are 15 compounds or groups of compounds for which SNARLs have been provided.

*Hazardous Air Pollutants*

The Clean Air Act requires EPA to regulate toxic air pollutants. Pollutants classified by EPA as hazardous are subject to regulations through National Emission Standards for Hazardous Air Pollutants (NESHAP). Potential airborne toxic chemicals are identified and prioritized in a stepwise process heavily based on both the toxicity of a chemical and the potential for public exposure. The highest priority is given to possible carcinogens, mutagens and teratogens. The list of 37 substances has been produced through such ranking of 632 organic chemicals initially identified by EPA as potentially hazardous pollutants.

The driving force behind actions leading to establishment of national environmental standards is the protection of the health and welfare of the US population as a whole. The decisions whether and what to regulate are reached centrally on the federal level. As exemplified above, pollutants are selected for regulation either because of high toxicity and extensive human exposure (trihalomethanes, hazardous air pollutants) or because of the outside pressure such as a law suit (ambient air lead, 129 priority pollutants). SNARLs are probably the only federal guidelines which are formulated specifically in response to requests on the state level. However, being derived solely on the basis of health-effect data, without regard for technological feasibility and costs (specifically required by the Safe Drinking Water Act) SNARLs are not always practical to use in major regulatory decisions. This centralized approach to regulating pollutants is highly justified when the health and welfare of the nation is at stake. However, it frequently does not suit the particular needs and priorities of individual states. The principal reasons are as follows:

(1) A relatively toxic chemical which is of low priority for regulators because of low human exposure on the national level may be manufactured and/or emitted in high quantities in one particular area of the country.

(2) The time factor involved in rulemaking by a federal agency is much too long to satisfy needs of state regulatory agencies for which prompt response to specific situations is the key to effectiveness.

(3) Different pressures – political, social, economic – act on the state and on the federal levels. These pressures will have a profound effect on the regulatory priorities. The very distance from the affected population will alter the regulator's perception of the urgency and magnitude of the problem.

Two examples from Massachusetts illustrate these discrepancies between EPA and the state regulatory agency.

*Chromium Emissions*

Massachusetts has a well-developed leather tanning industry which uses large quantities of chromium sulfate and other salts of trivalent chromium. Wastewater high in chromium compounds is the by-product of these industrial activities. The preferred method of disposal of this waste is treatment in a wastewater treatment facility followed by incineration of the resulting sludge and subsequent disposal of the incinerator ash by landfilling. During incineration, some chromium is emitted from a stack into the atmosphere. These air emissions, which are under the regulatory mandate of the Massachusetts DEQE, are the hardest to assess and regulate. Hexavalent chromium compounds are linked to human respiratory cancer. Because of that, there are separate occupational standards for different chromium compounds; metal and insoluble salts: 1 $mg/m^3$; chromic and chromous salts: 0.5 $mg/m^3$; non-carcinogenic hexavalent chromium compounds: 25 $\mu g/m^3$; carcinogenic hexavalent chromates: 1 $\mu g/m^3$.

According to the National Air Surveillance Network, background concentrations for total chromium in a typical urban environment ranged from 0.01 to 0.03 $\mu g/m^3$ in 1979 and was below 0.01 $\mu g/m^3$ in the Metropolitan Boston Area. Computer modeling of the emissions from the South Essex sludge incinerator in Salem, one of several such facilities in Massachusetts, indicated that the maximum excess concentration of total chromium in ambient air would be 0.0042 $\mu g/m^3$ at a downwind distance of 200 meters. It is unclear what percentage, if any, of this excess chromium concentration in air would be attributed to the potentially carcinogenic hexavalent form since hexavalent chromium is readily reduced in the environment to the trivalent form. Although under the optimal operating conditions of the incinerator no hexavalent chromium should be emitted, the worst case scenario was constructed where at one particular time 40% of the chromium emitted from the incinerator would be in the oxidized state. In that case, the total excess ambient air concentration of hexavalent chromium would be 0.0017 $\mu g/m^3$.

The Agency's task was to assess the acceptability of this excess chromium over background of about 0.01 $\mu g/m^3$. The final decision which allowed the South Essex incinerator to emit chromium at the levels predicted was made on the state level because the ambient air chromium is not regulated under the federal NAAQS or NIESHAP programs. It did not, however, solve the wider problem of regulating other sources. Presently there are ten large public treatment/incineration facilities either operating or under construction in Massachusetts which process chromium-containing wastewater. The amount of chromium entering these plants is not known but it is estimated that the combined ash produced daily in all the incinerators contains in the order of one thousand pounds of chromium. It is reasonable to assume that, with these large quantities of chromium in ash, there is the potential for significant air emissions of the metal. This is an urgent problem in Massachusetts which requires regulatory action in the immediate future.

*Tetrachloroethylene in Public Drinking Water Pipes*

In the Spring of 1980 it had been discovered that drinking water transported in vinyl-coated asbestos-cement (A–C) pipes in Massachusetts often contained elevated concentrations of tetrachloroethylene. The concentrations of tetrachloroethylene in the affected drinking water ranged from non-detectable to as high as 2,000 ppb in some instances. Tetrachloroethylene is used as a solvent during application of the vinyl coating to the asbestos-cement pipe during manufacture, and residual solvent leaches into water carried in these pipes.

There is no national drinking water standard (MCL) for tetrachloroethylene. In February of 1980, EPA issued SNARLs for 1-day, 10-day and chronic exposure to tetrachloroethylene of 2,300, 175 and 20 ppb, respectively. The SNARLs were based on the animal studies where exposure to large quantities of the substance produced organ toxicity. The guidelines were calculated by extrapolating the animal data to man with a large margin of safety (uncertainty factor of 100 for 1-day and 1,000 for 10-day and chronic exposure). Later, in response to the tetrachloroethylene contamination in the New England states, EPA Office of Drinking Water issued a temporary guideline, Suggested Action Guide (SAG) of 40 ppb. Most of the affected communities still exceeded the new EPA guideline of 40 ppb but at least the standard against which the purity of water was measured was relaxed. Public perception of the seriousness of any contamination of drinking water is such that there was a tremendous public pressure on DEQE to take an immediate action which would remove all traces of tetrachloroethylene or at least get it below the EPA guideline.

The problem in Massachusetts was as follows: if all water supplies where concentrations of tetrachloroethylene were in excess of 40 ppb were to be closed, 58 communities would have lost their source of tap water. On the other hand, if a guideline of, say, 100 ppb was used instead of 40 ppb, the problem would have been reduced to only 31 communities. That reduction would have meant more resources to handle the problem. SNARLs, and in this case SAG, are practically used as though they were legally enforceable national drinking water standards, which they are not. This is despite all the uncertainties involved in their computation and their lack of compliance with the Safe Drinking Water Act which specifically requires the economic and technologic realities to be considered in setting MCLs. This is because of the strong need for guidance among regulators on the state and municipal level. Thus, because of the regulatory weight given to such guidelines, it was not feasible to consider reevaluating the existing SNARL or SAG and possibly coming up with a guideline less stringent than that.

Recently, on March 4, 1982, EPA issued an advanced notice of proposed rulemaking for 14 volatile organic chemicals in drinking water, tetrachloroethylene being one of them. After considering the health effects, potential human exposure, treatment alternatives and costs of control and monitoring of the chemicals, EPA recommended an MCL for tetrachloroethylene in the 5–500 ppb range. This initial proposal is still subject to public comment, expert review and the Administrator's final decision, but it indicates that EPA considers a potential MCL much greater than 40 ppb. In the Spring of 1980, DEQE did not have the benefit of EPA guidance in this matter. It was apparent then that an MCL for tetrachloroethylene, if EPA were to develop the standard, would be probably higher than 40. However, unless such a standard was established through a rigorous regulatory process, with public participation, no state guideline less stringent than the EPA's 40 ppb could withstand the public pressure of the moment.

In time, the problem of contamination was eventually solved by DEQE through an aggressive program of periodic flushing of the pipes. Presently, it is either completely eliminated or under control in most of the affected communities.

The cases of chromium and tetrachloroethylene exemplify the many situations where regulatory decisions have to be made on the state level without the benefit of legally enforceable federal environmental standards. Most state agencies have a serious shortage of necessary professionals experienced in toxicology, risk analysis and standard setting. In the present political climate we cannot hope for improvement. Neither can the states count on increased, or even continuing, leadership from EPA with its own decimated budget and personnel.

If strong environmental programs are to survive on state levels, a climate of close cooperation between state regulatory environmental agencies must be created. Since each state has a different set of priorities, a network of cooperation should be flexible enough to accommodate the changing needs of the day. Thus, an interagency task force for ambient air standard for chromium would have different membership than that for drinking water guideline for an agricultural pesticide. The immediate advantage of such cooperation would be effective increase in human resources, less duplication of effort, broadened data base and more diversified approach to problems. In the long run, it would improve the ability of regulatory agencies to handle their particular regional situations and, ultimately, lead to strong environmental programs.

Dr. Halina S. Brown, Massachusetts Department of Environmental Quality Engineering, 1–11 Winter Street, Boston, MA 02108 (USA)

Safety Evaluation and Regulation of Chemicals. 1st Int. Conf., Boston, Mass., 1982, pp. 58–65 (Karger, Basel 1983)

# Federal Regulation of Toxic Substances in the Workplace

*Nicholas A. Ashford*

Center for Policy Alternatives, Massachusetts Institute of Technology, Cambridge, Mass., USA

The recognition of toxic substances in the workplace has implications not only for workers, but for the general population, since very often it is within the workplace setting that the effects of toxic substances are first recognized. The combination of high levels of exposure and an identifiable population at risk often yields the first clear hints of a health hazard. Examples are asbestos, vinyl chloride, and kepone.

The control of toxic substances in the workplace through federal legislation is effectuated through both the Occupational Safety and Health Act of 1970 and the Toxic Substances Control Act (TSCA) of 1976. The evolution of case law under the OSH Act has profoundly influenced other environmental legislation and promises to affect similarly the evolution of the TSCA. This paper will address the existing federal legislation, key court decisions interpreting the reach of the Occupational Safety and Health Act – especially with regard to the importance of undertaking risk assessments and cost-benefit analyses – and, finally, discuss the impact of the legislation on non-regulatory and private sector activities relevant to the control of toxic substances.

The OSH Act provides two general means of protection for workers: a statutory general duty of care and adherence to specific standards by employers. The Act imposes on virtually every employer in the private sector a general duty 'to furnish to each of his employees employment and a place of employment which are free from *recognized hazards* that are causing or are likely to cause death or serious physical harm ...'. A 'recognized hazard' may be a substance for which the likelihood of harm has been the subject of research giving rise to reasonable suspicion and for which an OSHA standard has not yet been promulgated. The burden of proving that a particular substance is a 'recognized hazard' and that industrial exposure to it is at a serious level is placed on OSHA. Since standard setting is a slow process, protection of workers through the employer's general duty obligation is especially important, but it is crucially dependent on the existence of good health effects data.

The OSH Act addresses specifically the subject of toxic materials. It states, under Section 6(b) of the Act, that the Secretary of Labor (Occupational Safety and Health Administration, OSHA) in promulgating standards dealing with toxic materials or harmful physical agents shall set the standard which 'most adequately assures, to the extent *feasible,* on the basis of the *best available evidence* that *no* employee will suffer material impairment of health or functional capacity even if such employee has a regular exposure to the hazard dealt with by such standard for the period of his working life.' By these words, one can see that the issue of exposure to toxic chemicals or carcinogens that have long latency periods is covered by the Act in specific terms.

Under Section 6(a) of the Act, without critical review, OSHA initially adopted as standards the 450 Threshold Limit Values (TLVs) recommended as guidelines from protection against the toxic effects of these materials. In addition, under section(6b) OSHA has set formal standards for a package of 14 carcinogens, asbestos, vinyl chloride, arsenic, DBCP, coke oven emissions, acrylonitrile, benzene, lead, and cotton dust. The burden of proving the hazardous nature of a substance is placed upon OSHA, as is the requirement that the proposed controls are technologically feasible. The necessarily slow and arduous task of setting standards substance-by-substance cannot realistically deal with 13,000 substances on the toxic substances list of the National Institute of Occupational Safety and Health (NIOSH) or NIOSH's list of 2,000 suspect carcinogens. Efforts had been made to streamline the process by proposing generic standards for carcinogens.

The inadequacy of the 450 TLVs set under Section 6(a) of the Act are legend. The TLVs had their origin as recommended guidelines by the American Conference of Governmental Industrial Hygienists (ACGIH) to protect the average worker from either recognized acute effects or easily recognizable chronic effects. The standards are based on animal toxicity data or the limited epidemiological evidence available at the time of the establishment of those guidelines. They do not address the sensitive populations within the work force, or those with prior exposure or existing disease; nor do they address the issues of carcinogenicity, mutagenicity, and teratogenicity. These standards were adopted en masse on May 21, 1971, as a part of the consensus standards that OSHA adopted along with those dealing primarily with safety. As an example of the inadequacy of protection offered by the TLVs, it might be remembered that the TLV for vinyl chloride was set at 200 ppm, while the new protective standard recommends no greater exposure than 1 ppm – a level recognized still to be unsafe, but all that the technology could deliver. Another example is the TLV for lead which was established at 200 $\mu g/m^3$, while the new lead standard required a level no greater than 50 $\mu g/m^3$, also recognizing that that level was not safe for all populations, e.g., pregnant women, or those with prior lead exposure.

Under Section 6(b) of the Act, new health standards dealing with toxic substances were to be established utilizing the mechanism of an open hearing and subject to review by the US Circuit Courts of Appeal. The evolution of case law associated with the handful of standards which OSHA promulgated through this section are worth considering in detail. The courts addressed the difficult issues of the adequacy of the scientific information necessary to sustain the requirement that the standards be supported by 'substantial evidence on the record as a whole'. The cases also addressed the extent to which economic factors were permitted or required to be considered in the setting of the standards, the meaning of 'feasibility', the question of whether a cost-benefit analysis was required or permitted, and, finally, the allowable reach of the OSH Act in addressing different degrees of risk.

In an early case challenging OSHA's authority to regulate 14 carcinogens, the Circuit Court of Appeals first addressed the issue of substantial evidence. For 8 of the 14 carcinogens there were no human epidemiological data. The industry challenged OSHA's ability to impose controls on employers in the absence of human data. Here the court expressed its view that some facts, such as the establishment of human carcinogenic risk from animal data, were on the 'frontiers of scientific knowledge', and that the requirement for standards to be supported by substantial evidence in these kinds of social policy decisions could not be subjected to the rigors of other kinds of factual determinations. Thus, OSHA was permitted to require protective action against substances known to produce cancer in animals but with no evidence of human carcinogenicity. It was not until much later that the Supreme Court in the benzene case placed limits on the reach of OSHA's policy determinations with regard to carcinogenic risk (see discussion below).

In the challenge to OSHA's asbestos standard – in which asbestos was regulated as a classical lung toxin and not as a carcinogen – the union challenged the laxity of the standard, claiming that OSHA improperly weighed economic considerations in its determination of feasibility. The union lost. OSHA indeed was permitted to consider economic factors in establishing feasibility. The court went on to say, however, that a standard might be feasible even if some employers were forced out of business, so long as the entire industry was not disrupted.

In the industry challenge to OSHA's regulation of vinyl chloride at 1 ppm, the Circuit Court of Appeals reiterated OSHA's ability to make policy judgments with regard to matters 'on the frontiers of scientific knowledge' when it declared that there could be no safe level for a carcinogen, and that since 1 ppm was the lowest feasible level, OSHA was permitted to force employers to comply even though it had performed no formal risk assessment or knew how many tumors would be prevented by the adoption of this protective level. Another noteworthy aspect of the case was the recognition that OSHA could act as a 'technology forcer' and require controls not yet fully developed at the time of the setting of the standard.

Following the first serious successful industry challenge of an OSHA standard on benzene in the Fifth Circuit Court of Appeals, the Supreme Court, in a controversial and divided majority opinion, chided OSHA for not having made an attempt to evaluate the benefits of changing the permissible exposure level for benzene from 10 ppm (the old TLV) to 1 ppm. The Court argued that OSHA is obligated to regulate only 'significant risks', and that without a [risk] assessment of some kind OSHA could not know whether the proposed control addressed a significant risk. The Court was careful to say that it was not attempting to 'statistically straitjacket' the agency, but that at a minimum the benefits of regulation needed to be addressed in order to meet the substantial evidence test. The Court did not give useful guidance as to what constituted a significant risk. It stated that a risk of death of 1 in $10^3$ was clearly unacceptable, while a risk of 1 in $10^6$ might be tolerated. This 3-order magnitude range, of course, represents the area that the arguments have always been centered on.

The implications of the benzene decision for future standards will depend on the nature of the particular OSHA administration. There is little question that had OSHA performed a risk assessment for benzene it could have argued that the risk it was attempting to address was in fact significant. On the other hand, a lax administration could just as easily decline to declare the risk as significant, and leave the protection at the old level of 10 ppm. The precise requirement and nature of a risk assessment sufficient to meet the substantial evidence test remains quite unclear.

Industry argued in the benzene case that not only must a risk assessment be performed, but a cost-benefit analysis must also be done, in which the risks of exposure must be balanced against the benefits of the chemical. The question, however, was not decided in the benzene case, but rather addressed in a later case challenging OSHA's cotton dust standard. There, the Supreme Court not only acknowledged that cotton dust did represent a significant risk, it also indicated that a cost-benefit balancing was neither required nor permitted by the OSH Act since Congress had already struck the balance heavily in favor of worker health and safety.

While it is clear that a cost-benefit analysis may not be the basis upon which the level of safety can be determined, nonetheless a cost-benefit analysis might be done in response to the requirements of Executive Orders of the President requiring economic impact analysis, or under the new Regulatory Reform Act of 1982. If these analyses are done, it may become impossible to determine whether the analysis influenced the level of safety required of employers in subsequent standards.

In the previous administration, OSHA attempted to promulgate a generic cancer standard by which questions of science policy as law, already settled in cases dealing with other standards, would be codified in a set of principles. Among the important scientific issues 'resolved' by the OSHA generic standard are:

(1) Positive results in human epidemiological studies alone, or positive results in at least one animal bioassay meeting certain minimum standards of quality and in the presence of supporting evidence, will suffice to establish the 'inference' that a substance is a carcinogen.

(2) Negative results in animal bioassays or human epidemiological studies will not be considered in deciding whether a chemical is a carcinogen if positive results have been obtained in either an animal bioassay or epidemiological study, unless the negative studies meet stringent quality standards.

(3) In animal bioassays, any tumor produced (including benign tumors), or an increase in the incidence of tumors which occur spontaneously, or the occurrence of tumors at the site of application of a substance by inhalation or oral routes, will establish the inference that the substance is a carcinogen.

(4) Tumors induced in animals at high dosage levels (including the maximum tolerated dose) will be considered evidence of carcinogenicity of that substance, unless strong evidence can be supplied that the metabolism of the substance in question is such that at low dose, no carcinogenic metabolites are produced.

These principles seem to represent a reasonable consensus of the state-of-the-art in carcinogenicity testing. The resolution of these issues once and for all in court could, if it occurs, greatly reduce the time and resources required of OSHA in addressing these topics during regulatory proceedings. It is not yet clear to what extent the remaining allowable areas for discussion would result in lengthy and protracted regulatory proceedings. Besides the remaining scientific issues related to the determination of carcinogenicity, OSHA would still be required to determine what level of control is feasible, and this is likely to remain a lengthy and difficult process, subject to delay by court challenge.

During the process of developing the proposed generic standards, OSHA and NIOSH developed lists of chemical substances likely to be classified as Class I suspect carcinogens. Each agency arrived at a list of about 250 substances. It is thus clear that some method is necessary for expediting the process by which standards are set if a reasonable effort is to be made to protect workers against exposures to a large proportion of all the known hazards. Even with the streamlined procedure, it is not likely that this can be accomplished for many years. On the other hand, large numbers of workers are exposed to only a few of these substances, and the publication of OSHA's priority lists will serve as warning to employers and employees alike that certain substances are thought to present serious hazards and are likely to be regulated in the near future. Such warnings can be expected to result in voluntary efforts to reduce exposure, and this voluntary effort will probably be an important part of the total 'benefits', in terms of reduced exposure, which will result from the generic carcinogen standard.

The proposed generic cancer standard was somewhat revised by the prior administration when the requirement that OSHA standards must address 'significant risk' was added by the Supreme Court in the benzene case. The present administration is reconsidering the policy entirely.

The implementation of the OSH Act with the burden of proof placed on the government cannot keep up with the proliferation of chemicals into the work environment, and it is certainly no safeguard against general environmental contamination or consumer product hazards. The burden of proving 'unreasonable risk' in consumer products is likewise placed on the Consumer Product Safety Commission, which has taken action on only a few chemical problems, including fluorocarbons, TRIS®, vinyl chloride, and formaldehyde in foam insulation. The need for legislation which could exercise control over all activities traceable to the manufacturing, use, and disposal of chemicals by shifting the burden of proof to the producer/user was apparent soon after the OSH Act and the Consumer Product Safety Act were passed. After 5 years of intense effort, a coalition of traditional environmental groups and the labor movement succeeded in getting TSCA enacted. This coalition is noteworthy because labor, fearing earlier that one could not have environmental quality without an adverse impact on the economy, no longer felt itself seriously pressured by 'environmental blackmail'.

TSCA enables the Environmental Protection Agency (EPA) to require data *from industry* on the production, use, and health/environmental effects of chemicals. EPA may regulate by requiring labeling, by setting tolerances, or by banning completely and requiring repurchase or recall. EPA may also order a specific change in chemical process technology. In addition, TSCA gives aggrieved parties, including consumers and workers, specific rights to sue for damages under the Act, with the possibility of awards for attorneys' fees. This feature was missing under the OSH Act.

The strength of the TSCA is in its ability to shift onto the producer the requirement to prove safety to the extent that exposure to the substance does not present an 'unreasonable risk of injury to human health or the environment'. Used together, these two pieces of legislation, the OSH Act and TSCA, provide potentially comprehensive and effective information generation and standard-setting authority to protect workers. In particular, the information generation activities under TSCA can provide the necessary data to have a substance qualify as a 'recognized hazard' which, even in the absence of specific OSHA standards, must be controlled in some way by the employer in order to meet the general duty obligation under the OSH Act to provide a safe and healthful workplace.

In addition to the legislative strictures provided by OSHA and TSCA, the legislation has had an impact on nonregulatory and private sector activities relevant to the control of toxic substances. Through recent regulations promulgated by OSHA giving workers access rights to exposure and medical examination information, and through recent decisions of the National Labor Relations Board under the National Labor Relations Act which give unions the right to chemical exposure information, worker self-help activities are likely to increase substantially. These involve not only the traditional filing of worker compensation claims, but also collective bargaining activities for unionized workers. Information is indispensable for use in tort and product liability suits against the manufacturers of toxic substances, such as the proliferation of suits seen in the asbestos industry. Under both the OSH Act and the National Labor Relations Act, workers have the right to refuse hazardous work, and the mere existence of standards – even without their enforcement – establishes the legal argument that work with a particular substance might be hazardous and hence workers are entitled to self-protection by refusing dangerous tasks. Thus, the impacts of risk assessment and hazard identification information, if made available to workers and unions, will result in a predictable increase in private sector solutions to problems inadequately addressed by regulation.

Nicholas A. Ashford, PhD, JD, Massachusetts Institute of Technology,
Center for Policy Alternatives, E40-250, Cambridge, MA 02139 (USA)

Safety Evaluation and Regulation of Chemicals. 1st Int. Conf., Boston, Mass., 1982, pp. 66–76 (Karger, Basel 1983)

# Conventional Approaches to Safety Evaluation

*Carolyn T. Miller, Daniel Krewski, Ian C. Munro*

Health Protection Branch, Health and Welfare Canada, Ottawa, Canada

Of the realities involved in chemical safety evaluation, one of the most fundamental must surely be uncertainty: uncertainty with respect to the causal factors underlying the diseases we wish to prevent; uncertainty in the predictive value of the procedures used to detect hazardous materials; and uncertainty as to the social values to be placed on the various risks and benefits associated with chemical usage. The challenge of safety evaluation, therefore, is to use what knowledge *is* available to shrink the degree of uncertainty, and thus make the best possible decisions in the face of uncertainty.

The process of safety evaluation may be described in terms of four phases [43]. Initially, risk *identification* involves demonstrating that a well-defined hazard exists. Once a potential risk factor has been identified, quantitative risk *estimation* techniques may, in principle, be used to determine the magnitude of the risk involved. Next, risk *evaluation* attempts to project the potential impact of alternative control options with respect to risks, costs and benefits as they might affect each of the sectors of society involved. Finally, strategies for the *management* of risk may be developed to control or distribute risks and benefits in the manner perceived as optimal. An important component of the risk management phase is monitoring of the actual impact of the control measures once they have been implemented.

It seems generally accepted that scientists share the major responsibility for the identification and estimation of risks. Responsibility for regulations, on the other hand, rests largely with the legislators and regulatory agencies involved. Between the science of risk estimation and the practicalities of risk management, however, an interface must be built so as to permit socially responsive risk evaluation.

*Risk Identification*

The four major methods for identifying chemical carcinogens are the analysis of physical and chemical properties of the test compound; short-term tests in model systems representing specific components of biological response mechanisms; long-term chronic bioassays in laboratory animals; and epidemiological studies of human populations. The utility of each of these methods has recently been discussed by the Office of Technology Assessment of the US Congress [29]. For brevity, we will restrict our attention here to the latter two approaches, as these presently provide the most direct evidence of potential human health hazards.

*Epidemiological Studies of Human Populations*

Epidemiological studies may be designed to explore two distinct types of risk factors [10]. The first includes general environmental conditions such as diet and life-style that can only be inferred indirectly from geographical and secular changes in cancer incidence, and from observations in migrant populations. The second includes more specific risk factors which may be identified with some certainty and linked to affected and unaffected individuals.

General risk factors for cancer have been explored through analysis of national registries of cancer incidence and mortality. The Bureau of Epidemiology of Health & Welfare Canada estimates that cancer accounts for 19 and 22% of all deaths among males and females, respectively [2]. Recent attempts to discern the proportion of human cancer attributable to man-made chemicals suggest that such chemicals may in fact contribute relatively little to the total burden of disease [10, 16, 17]. Working with United States statistics, *Doll and Peto* [10] in fact considered environmental contamination as a possible causal agent for only about 2% of the total cancer burden, while occupational exposure was estimated as being responsible for 2–8%. Other estimates, however, have been somewhat higher [13].

To more clearly elucidate the causal role of specific risk factors, epidemiological studies may involve clinical observations on relatively small patient populations, or retrospective and prospective studies of much larger populations. In the case of chemically-induced bladder cancer, for example, all of these methods in sequence have proven useful in the risk assessment process [3].

Generally, one can expect definitive results only if exposures are high and affect a well-defined subpopulation. Further, association of exposure and disease will be facilitated if the end point is unusual or rare. Accurate diagnoses, precise measurements of exposure, rapid onset of disease and knowledge of confounding risk factors are all criteria which, although they would simplify the interpretation of epidemiological data, are all too infrequently encountered in practice [1].

Perhaps of most importance is the fact that epidemiology deals with an assessment of disease etiology after the fact. By the time a carcinogenic effect becomes detectable, exposure may have occurred over a period of 15 or more years. Dependence on epidemiology as the primary tool for hazard identification is therefore inconsistent with a preventive regulatory philosophy.

*Toxicological Studies in Animal Models for Man*

Conventional methods of risk assessment have relied heavily on mammalian toxicology. For chemicals with direct routes of exposure in man, such as drugs, food additives and pesticides, an extensive toxicological profile must be generated to provide reasonable assurance of safety prior to licensing or registration [11, 12, 15]. In Canada, the composition of the test battery is flexible, and can be adjusted according to the complexity of individual cases. However, guidelines are being drafted to provide a general indication of the tests considered appropriate. An issue of current debate is the need for a minimum premarket data package for all chemicals not covered by existing product-specific legislation. The European Economic Community Council Directive of September 18, 1979, in fact specified a number of premarket requirements for the sale or distribution of any chemical where production exceeds a specified limit [7].

Many sets of guidelines for the performance and evaluation of cancer bioassays have been published by regulatory authorities, with significant comment by a variety of individual experts in the field [4, 25, 31, 36, 37, 42]. Although differences of opinion remain with respect to certain criteria such as the number and strength of doses to be tested, the duration of the study and the extent of histopathological examination, the basic principles of the bioassay are held largely in common. Differences may reflect the particular objective of the study, which may be directed towards screening, dose extrapolation, or the elucidation of mechanisms of action.

Often, the regulatory toxicologist who must review the resulting safety data has had no hand in the design, execution or reporting of the study, these being the responsibility of the manufacturer or distributor of the chemical. However, current initiatives such as the guidelines being prepared by the

Organization for Economic Cooperation and Development may make experimental design less of a variable, and ensure the application of sound toxicological design principles. Further, the introduction in 1977 of Good Laboratory Practices legislation in the United States [14], and the related Memorandum of Understanding between Canada and the US, now provide considerably more access to and control of the in-life performance of toxicological studies than has been the case in the past.

*Risk Estimation*

Toxicological evaluation begins with validation of the study. This phase involves an appraisal of the study design, execution, and reporting to determine whether the study is adequate to elucidate the toxic phenomena of concern [28]. Ideally, the toxicologist would have access to all of the raw data including original records of observations during the in-life portion of the study, although rarely has toxicological evaluation included such full validation. Nonetheless, sufficient detail must be appraised to permit validation of the doses actually administered, the effective number of animals adequately assessed, adherence to the protocol and accuracy of reporting. Any reservations regarding the validity of the study must be clearly enunciated, and considered in the subsequent interpretation of thc data.

If it is determined that a biologically important effect has been induced and that the effect is not likely to have occurred by chance alone, it can be concluded that exposure to the chemical may be associated with a health hazard, at least in qualitative terms. The next step in the safety evaluation process requires a quantitative estimate of the magnitude of the risk to man.

*Safety Factors and Mathematical Models*

A common approach to risk estimation would first identify the most sensitive parameter in the most sensitive species. The highest dose inducing *no* such effects would than be determined [40, 41]. Taking this 'no observed effect level' (NOEL) at face value – namely that no *observed* effect really does indicate *no* effect – and allowing a margin of error through the use of a safety factor to encompass the possibility that man may be more sensitive than the test animal species tested and that individuals in the human population may be of varying sensitivity themselves, an acceptable daily intake (ADI) is derived. The magnitude of the safety factor is often 100-fold [23] but may be smaller if practicalities impose themselves, or larger if special evidence indicates the need for added prudence.

Precedent has indicated that for most detrimental effects on biological systems, this safety factor (SF) approach is reasonable. Most mammalian tissues appear to have in common natural defense mechanisms that allow them to tolerate a certain level of exposure to many chemicals. These defense mechanisms include metabolic detoxification, chemical neutralization of ions and radicals, tissue regeneration, and immunological surveillance. Fundamental to the resilience of biological systems to chemical exposure is their redundancy. Chemical damage to a small number of liver or kidney cells, for example, may be tolerated because organ function is maintained by residual normal tissue until regenerative cycles have replaced the damaged cells.

Mutation of the genetic material of cells, however, does not have this form of biological resilience. Genetic damage is transmitted to daughter cells, and a single aberrant cell could give rise to a malignant clone regardless of the huge excess of surviving normal tissue. Thus, the utility of the safety factor approach in the risk assessment of carcinogens has been seriously questioned [5, 34, 39].

The underlying assumption in the NOEL/SF procedure is that a threshold exists in the dose-response relationship between the NOEL and the lowest effective dose. It is now widely recognized, however, that the NOEL depends on the number of animals tested and that the observation of no responses in even a relatively large group of test animals does not provide absolute assurance of safety. With an agent tested at a level at which the risk in the entire population is actually 1%, for example, there is a better than ever chance of observing no effects in a group of 50 animals exposed to that dose [6].

Another approach to risk estimation involves extrapolation via mathematical models fitted to the observed dose-response data [21]. Because of the statistical difficulties involved in establishing a threshold in the dose-response curve on purely empirical grounds [34], however, most mathematical models have dispensed with the threshold concept. Although any nonzero dose will be associated with some increase in risk in the absence of a threshold, a virtually safe dose (VSD) corresponding to some suitably low level of risk may still be estimated [5].

Although many different models have been proposed, all predict linear, sublinear or supralinear behavior of the dose-response curve in the low-dose region [26]. Experience has shown that the data obtained from studies involving even relatively large numbers of animals may fit more than one of these model types equally well [22]. Therefore, mathematical models per se do not assist in predicting the low-dose response unless there is some independent knowledge of mechanism to direct the selection of a model. If, for example, the test compound acts in a manner that is dose-wise additive to an ongoing process, then it can be shown that the dose-response curve will in fact be linear in the low-dose region [9]. In this case, the safety factor approach may seriously underestimate the actual risk at low doses and lead to an ADI which could be substantially greater than the VSD.

*Exposure Assessment*

A crucial component of risk estimation is an assessment of the exposure likely to occur in man. With pesticides, for example, residue limits for each crop are experimentally determined. These residues reflect the mode and rate of application required for efficacy with respect to pest control. By applying these residue limits to food consumption figures, estimates of chemical exposure via the oral route from each crop on which the compound is used may be made. The contributions of existing uses and proposed uses are combined to get a theoretical daily intake (TDI). If the TDI for man falls below the ADI based on toxicological data, use of the pesticide would be recommended for further consideration.

*Mechanisms of Action*

Practical decision making would clearly be improved by the application of all of the biological tools available for the reduction of uncertainty. With respect to cancer, consideration should be given to the dose actually delivered to the target tissue and to nonspecific stress and its potential for enhancement of carcinogenesis [4]. Factors involved in the last instance could include the occurrence of secondary events which, although not a part of the causal sequence leading to cancer, are induced in the treated animals. More convincing would be the occurrence of antecedent events believed to represent stages in malignant transformation, such as enzyme-altered foci in liver. In the liver, a number of histochemically definable changes generally precede the occurrence of hepatocellular carcinoma. Demonstration of a clear threshold in these histochemical changes would lend credence to a threshold for tumorigenesis in the rodent model.

*Interspecies Extrapolation*

A fundamental consideration with respect to the relevance of animal models to human safety evaluation is the qualitative or even quantitative relationship between effective doses and target sites in animals and those in man [3, 18–20, 30, 31, 35, 38]. A recent report on selected cancer bioassay data reviewed by the Meselson Committee [24] suggested that differences exist in both potency and target site between test animals and man. However, most known human carcinogens have also been shown to affect one or more animal species, although not always at the same site [27].

Information on mechanisms of action can be particularly useful in trans-species extrapolation if the indicator events can be measured with technical reproducibility at low doses and in man himself. Procedures applicable in man include screening of urine for mutagens, lymphocytes for chromosome damage or DNA repair, sperm for various anomalies, and hemoglobin for alkylation. The more directly related the indicator event is to the postulated steps in carcinogenesis, the more weight can be assigned to the results. Other factors that could be observed to further elucidate the operative mechanisms would include hormonal activity, effects on immune response, and synergism with known inducers.

### *Risk Evaluation*

The estimation of carcinogenic risk involves a careful appraisal of biological and statistical factors arising from animal bioassays. This risk is then evaluated along with the best available estimates of potential human exposure, taking into account intensity, duration, and frequency, as well as the population at risk. At this point, the regulatory process really has just begun. If risks cannot be reduced to an acceptable level without adverse health or economic effects on consumers, workers or industry, then it seems clear that a responsible and equitable regulatory system should provide for a weighing of health and other benefits against risk [8, 20, 21, 32, 33, 44].

The benefits of chemical use may be considered in terms of improved health, economic gain, and the aesthetic qualities of appeal and convenience. Health benefits include the provision of adequate supplies of nutritious food through control of spoilage organisms or avoidance of disease, as in the control of hospital infections through antiseptic techniques using hazardous chemicals. Economic gains may accrue to the manufacturers and distributors of agrichemicals through business expansion and product development, and to farmers through increased productivity and higher quality produce. The convenience of long shelf life is a major attribute of food preservatives, while product appeal is achieved with colours and flavours.

*Risk Management*

Each available option for risk management, ranging from public education to restricted use to outright prohibition may be associated with a different distribution of risks and benefits, as well as a different sum total of risks and benefits. Other factors involved in deciding upon a preferred option might include the following [26]: availability of alternatives; feasibility of enforcement; impact on trade; consumer expectations; impact on future decisions.

In Canada, the regulatory agencies are responsive to the views of diverse sectors of society, through ad hoc advisory panels of experts in appropriate medical, nutritional or technical fields. Certain consumer associations are also represented on these panels. In addition, the agency has the responsibility to inform consumers of potential risks through product labelling and educational programs.

Further, we are currently committed to revoke any unnecessary regulations which impose costs on industry without appropriate benefits. A Treasury Board Directive of 1979 requires a detailed socio-economic impact statement for any proposed action related to health, fraud or safety where the impact on industry may exceed twelve million dollars annually.

*Improving the Risk Assessment Process*

Two general strategies for improving the risk assessment process can be discerned. The first involves increasing the precision of the estimates of exposure, risk and benefit through more detailed research. Recognizing the endless possibilities for refining such estimates, the limitations of such a strategy become apparent. Operating alone, this research strategy would require that we make an enormous commitment of resources, with no guarantee that such efforts would ultimately provide satisfactory results.

As with our formal legal and judicial systems, chemical safety evaluation often involves the study of precedent. Thus, the second strategy recognizes the uncertainty arising from our current ignorance of disease mechanisms and puts limits on the testing resources available. This strategy in effect defines the past and present approaches employed in the management of hazardous chemicals. With this precedent as a starting point, incremental changes in risk management may then be introduced. The role of science in this strategy would be to provide a cogent summary of existing precedent in order to generate a perception of the need for change.

Even in the face of uncertainty regarding the magnitude of the risks involved, the precedent strategy at least achieves a decision. It may be an unduly liberal decision if past judgements have in fact been wrong. The advantage of this strategy, however, is that it allows, indeed it almost guarantees, the accumulation of societally acceptable policies in the long run.

### *Conclusions*

To be workable, the preceding strategy for improving the risk assessment process imposes responsibilities on both the regulator and the regulated. The regulator must be able to use existing knowledge to minimize the burden of regulatory research and at the same time maintain a preventive capacity. The regulated must use scientifically sound principles of selection, design and execution of compound-specific research to provide a sound basis for decision making. The regulatory agency must be fully cognizant of, and make prudent use of, the precedent in the field. Further, it must be sensitive and responsive to the value judgements of the society it serves and choose options that are broadly acceptable.

To achieve all this, there must be open communication among the societal sectors involved. Where there is uncertainty, there must be room for debate among those best equipped in knowledge and experience to choose acceptable courses of action. Hopefully, the present ground swell of public interest in chemical hazards will be used to contribute constructively to the complex social issue of chemical safety evaluation. Perhaps one day we may actually achieve the accord of industry, consumers, workers and the government. Only then will safety evaluation become 'conventional'.

### *References*

1 Breslow, N.E.; Day, N.E.: Statistical methods in cancer research, vol. 1. The analysis of case control studies. IARC Scientific Publications, No. 32 (International Agency for Research on Cancer, Lyon 1980).

2 Health and Welfare Canada: Cancer patterns in Canada 1931–1974 (Bureau of Epidemiology, Health and Welfare Canada, Ottawa 1979).

3 Clayson, D.B.: Relationships between laboratory and human studies. J. envir. Path. Toxicol. *1:* 31–40 (1977).

4 Clayson, D.B.: Carcinogens and carcinogenic enhancers. Mutation Res. *86:* 217–229 (1981).

5 Cornfield, J.: Carcinogenic risk assessment. Science *198:* 693–699 (1977).

6 Cornfield, J.; Carlborg, F.W.; Van Ryzin, J.: Setting tolerances on the basis of mathematical treatment of dose-response data extrapolated to low doses; in Plaa, Duncan, Proc. 1st Int. Congr. Toxicology, pp. 143–164 (Academic Press, New York 1978).

7 Council of European Communities: Directive of 18 September 1979. Sixth amendment to directive 67/548/EEC (Council of European Communities, Brussels 1979).

8 Crouch, G.; Wilson, R.: Risk/benefit analysis (Ballinger, Massachusetts 1982).

9 Crump, K.; Hoel, D.; Langley, C.; Peto, R.: Fundamental carcinogenic processes and their implications for low dose risk assessment. Cancer Res. *36:* 2973–2979 (1976).

10 Doll, R.; Peto, R.: The causes of cancer. Quantitative estimates of avoidable risks of cancer in the United States today. J. natn. Cancer Inst. *66:* 1193–1308 (1981).

11 Environmental Protection Agency: Pesticide programs; proposed guidelines for registering pesticides in the United States; hazard evaluation: humans and domestic animals. Fed. Reg. *43:* 37336–37403 (1978).

12 Environmental Protection Agency: A handbook of key federal regulations and criteria for multimedia environmental control (Environmental Protection Agency, Washington 1979).

13 Epstein, S.: The politics of cancer (Anchor Press, Garden City 1979).

14 Food and Drug Administration: Good laboratory practices for nonclinical laboratory studies. Fed. Reg. *43:* 59982–60004 (1979).

15 Food Safety Council: Proposed system for food safety assessment. Fd Cosmet. Toxicol. *16:* suppl. 2, pp. 97–108 (1980).

16 Higginson, J.: Multiplicity of factors involved in cancer patterns and trends. J. envir. Path. Toxicol. *3:* 113–125 (1980).

17 Higginson, J.: Importance of environmental and occupational factors in cancer. J. Toxicol. envir. Hlth *6:* 941–952 (1980).

18 IARC: Chemicals with sufficient evidence of carcinogenicity in experimental animals – IARC Monographs, vol. 1–17. IARC International Technical Report 78/003, Lyon 1980, pp. 21–84.

19 IARC: An evaluation of chemicals and industrial processes associated with cancer in humans based on human and animal data. IARC Monographs, vol. 1–20. Cancer Res. *40:* 1–12 (1980).

20 Interagency Regulatory Liaison Group: Scientific bases for the identification of potential carcinogens and estimation of risks. A. Rev. Publ. Hlth *1:* 341 (1980).

21 Krewski, D.; Brown, C.: Carcinogenic risk assessment. A guide to the literature. Biometrics *37:* 353–366 (1981).

22 Krewski, D.; Van Ryzin, J.: Dose response models for quantal response toxicity data; in Csorgo, Dawson, Rao, Saleh, Statistics and related topics, pp. 201–231 (North Holland, Amsterdam 1981).

23 Lehman, A.J.; Fitzhugh, O.G.: 100-fold margin of safety. Ass. Fed. Drug Officials Bull. *18:* 33 (1954).

24 Meselson, M.S.; et al.: In Contemporary test control practices and prospects. The Report of the Executive Committee, National Academy of Science, vol. 1, pp. 75–81 (National Academy Press, Washington 1975).

25 Munro, I.: Considerations in chronic toxicity testing. The chemical, the dose, the design. J. envir. Path. Toxicol. *1:* 183–197 (1977).

26 Munro, I.; Krewski, D.: Risk assessment and regulatory decision making. Fd Cosmet. Toxicol. *19:* 549–560 (1981).

27 Munro, I.C.; Krewski, D.R.: Regulatory considerations in risk management; in Stich, Carcinogens and mutagens in the environment, vol. II, Naturally occurring products (CRC Press, Boca Raton, in press).

28 Office of Pesticide Programs, US Environmental Protection Agency and Health Protection Branch, Health and Welfare Canada: Guide to pesticide registrants for audit of IBT studies (Health & Welfare Canada, Ottawa 1978).

29 Office of Technology Assessment: Assessment of technolgies for determining cancer risks from the environment (US Government Printing Office, Washington 1981).

30 Pitot, H.C.: Relationships of bioassay data on chemicals to their toxic and carcinogenic risk for humans. J. envir. Path. Toxicol. *3:* 431–450 (1980).

31 Purchase, I.F.H.: Inter-species comparison of carcinogenicity. Br. J. Cancer *41:* 454–468 (1980).

32 Rall, D.P.: Issues in the determination of acceptable risk. Ann. N.Y. Acad. Sci. *363:* 139–144 (1981).

33 Roderick, H.: An alternative to risk-benefit analysis in government decision-making about chemical carcinogenesis. Ann. N.Y. Acad. Sci. *363:* 233–244 (1981).

34 Schneiderman, M.A.; Decouflé, P.; Brown, C.C.: Thresholds for environmental cancer. Biologic and statistical considerations. Ann. N.Y. Acad. Sci. *329:* 92–130 (1979).

35 Shubik, P.; Clayson, D.B.: Applications of the results of carcinogen bioassays to man. IARC Sci. Publ. *13:* 241–252 (1976).

36 Sontag, J.A.; Page, N.P.; Saffiotti, U.: Guidelines for carcinogen bioassay in small rodents. DHEW Publication No. (NIH) 76–801 (Department of Health, Education & Welfare, Washington 1976).

37 Tomatis, L.: The value of long-term testing for the implementation of primary prevention; in Hiatt, Watson, Winsten, Origins of human cancer (Cold Spring Harbor Laboratory, Cold Spring Harbor 1977).

38 Tomatis, L.: The predictive value of rodent carcinogenicity tests in the evaluations of human risks. A. Rev. Pharmacol. Toxicol. *19:* 511–530 (1979).

39 Truhaut, R.: An overview of the problem of thresholds for chemical carcinogens; in Davis, Rosenfeld, Carcinogenic risks/strategies for intervention. IARC Scientific Publication No. 25, pp. 191–202 (IARC, Lyon 1979).

40 Vettorazzi, G.: Handbook of International Food Regulatory Toxicology (CRC Press, Boca Raton 1980).

41 Weil, C.S.: Statistics vs safety factors and scientific judgement in the evaluation of safety for man. Toxicol. appl. Pharmacol. *21:* 454–463 (1972).

42 Weisburger, J.H.; Williams, G.M.: Carcinogen testing. Current problems and new approaches. Science *214:* 401–407 (1981).

43 Whyte, A.; Burton, I.: Environmental risk assessment (Wiley, New York 1980).

44 Wilson, R.: Risk/benefit analysis for toxic chemicals. Ecotoxicol. envir. Safety *4:* 370–383 (1980).

C.T. Miller, PhD, Health and Welfare Canada, Health Protection Branch,
Tunney's Pasture, Ottawa, Ont. K1A 0L2 (Canada)

Safety Evaluation and Regulation of Chemicals. 1st Int. Conf., Boston, Mass., 1982, pp. 77–84 (Karger, Basel 1983)

# Biological Bases for Risk Assessment

*Joseph V. Rodricks*[a], *Robert G. Tardiff*[b]

[a] Environ Corporation, and [b] National Academy of Sciences, Washington, D.C., USA

This paper will examine two of the three components of risk assessment and focus on the extent of the biological data base that supports some of the inferences and judgments necessary to make such assessments. The authors take as a major premise that the products of risk assessment are but one aspect of the decision-making processes of government and industry. Determining the nature and extent of the human risks from chemical exposures that can be tolerated in different circumstances involves a set of policy judgments that go beyond risk assessment. Such decisions usually incorporate the analysis of benefits and costs as well as risks. By contrast, the role of the risk analyst is to convey properly to the decision-maker the nature and probable magnitude of the risk and the many uncertainties associated with its assessment.

Before proceeding further with the problems of performing a risk assessment, it is useful to define the terms of discussion. There are as yet no universally accepted definitions. The following shall be used in this paper:

(A) *Hazard evaluation* is the evaluation of scientific data for purposes of characterizing the *nature* and *severity* of the health hazards that may be associated with exposure to a chemical substance. It is based on analysis of data gained from epidemiological, clinical, animal, and in vitro studies, and its goal is to characterize the toxicological effects of a substance.

(B) *Exposure assessment* is the analysis of scientific data for purposes of identifying the size and nature of the human population that may be exposed to a chemical substance and estimating the frequency, duration, and level of exposure.

(C) *Quantitative risk estimation* is the evaluation of both hazard and exposure information for purposes of estimating the likelihood that the hazards associated with a substance will be realized in the human population group at potential risk under their conditions of exposure. Such estimations have usually taken one of the following forms: (i) Estimation of the margin-of-safety associated with the conditions of exposure of the human population group at potential risk. This approach has been taken for noncarcinogens. While it does not involve an explicit probability estimate, the magnitude of the safety margin provides an indirect measure of risk. (ii) Estimation of the mathematical probability that a specific hazard will be realized under the conditions of exposure of the human population group at potential risk. This approach thus far has been applied only to carcinogens.

Risk assessment involves bringing together all of the above information and analyses, the qualitative as well as the quantitative, to yield a complete characterization of the risks and a statement about the uncertainties in the assessment.

## *Analysis of Concepts*

Our comments in this presentation will be directed to the hazard evaluation and quantitative risk estimation phases of risk assessment. We shall focus on three major concepts: first, the scientific basis for defining an effect as potentially hazardous or as 'adverse'; second, the biological basis for qualitative interspecies extrapolation; and, third, the toxicological underpinnings for biological thresholds and for quantitative extrapolation.

### *'Adverse' Effects: Biological Criteria*

There is virtually no disagreement that clinically diagnosed and frank illness or injury in humans, such as kidney failure, brain degeneration, and sterility, are indeed adverse. However, when small biological changes are observed in humans (for example, slight increases in the activity of an enzyme) with no measurable alteration in organ function or when similar observations are made in laboratory animals, substantial controversy surrounds the classification of such effects as 'adverse'. The differences in interpretation often relate to the prospect that the observed effect reflects either a compensatory response to stress, or the likelihood that the biochemical indicator does not constitute a determinative step in the alteration of organ function. Because the presence of change alone is not necessarily indicative of pathological alteration, additional experimental evidence to clarify the possible role in pathogenesis of the altered biochemical step is obviously desirable. It should be recognized, however, that for many changes, there are no readily identifiable experimental means for determining the ultimate health consequence. In such cases, a policy decision may have to be made whether to treat such effects as adverse.

Increased certainty that effects are indeed adverse can be acquired if two experimental conditions are observed: first, an increase in severity of effect with increasing dosage in the continuum of the dose-response relationship, and second, concordant observations on the following: gross characteristics of the intact animals, discrete effects at the organ-system level, and distinct alterations at the cellular and subcellular levels.

It must be recognized that circumstances may influence the categorization of an effect as either adverse or nonadverse. For example, the proliferation of the smooth endoplasmic reticulum with concomitant increase in the specific activity of some mixed function oxidases may be adaptive and protective if there is a relative increase in detoxification pathways, but may be considered deleterious if there is a relative increase in the enzyme systems that activate foreign compounds to more toxic metabolites.

*Interspecies Extrapolation: Comparative Biology*

It is generally recognized that for the assessment of human risks to most chemical exposures, only data from laboratory animals will be available. Three considerations enter in the use of laboratory animals for describing toxic hazards. These considerations are the ethics of human testing, the practicality of using small animals, and, most importantly, the unity of biology. Clearly, humans and laboratory rodents have in common many fundamental genetic, physiologic, and biochemical processes. Similarities include DNA as the source of all genetic information, the general reproductive cycle, functions of all major organs, and general biochemical processes for metabolism. For all of the broad similarities, there are many specific dissimilarities which are the source of varying degrees of uncertainty in drawing inferences about effects in one species based on data collected in another. For example, humans and rats have different numbers of genes, not all of which have similar functions. Rates of enzyme reactions are often different from species to species. In some cases one species has an enzymic pathway that is absent in another. Size and caloric needs are often different.

The risk analyst is faced with such dilemmas in the evaluation of composite data on a particular substance. If the data are concordant across all species tested and with consistent dose-response characteristics in each species, then the level of certainty that the substance is a potential toxicant in humans is quite high, a situation not often achieved. If the data appear discordant[1], some analysts have used the rule of thumb of selecting the experimental species in which toxic effects appear at the lowest dose, and of assuming that those pathologic endpoints are potentially referable to humans. This procedure involves the injudicious practice of discarding other data, when no scientific grounds exist for doing so. Some probability of error exists in such a procedure; however, since the false-positive and false-negative rates for conventional toxicity tests are essentially undefined, there can be no precise quantitative estimate of the uncertainty.

Over the past several years, there has been increased interest in selecting data from an experimental species that biotransforms the test substance in the same manner as does man. This biologic base has fostered an increase in metabolism studies; however, most such studies have been conducted in laboratory species only, with very little corresponding data in humans. Thus, the dilemma remains, despite the generation of new data. It is further compounded by the fact that, to be truly useful, metabolism studies need to be sufficiently detailed to reveal the nature of the toxic metabolite(s) and its target site concentration. These are exceedingly difficult technical tasks. There still remains the possibility of utilizing human cells in culture to obtain biotransformation data that could be compared to similar data from laboratory animals; and for some substances, the possibility also exists of in vivo human studies with very small quantities of compounds labeled with stable isotopes. While such information would be of substantial assistance in determining the most appropriate species to function as a surrogate for humans, the choices are unlikely to be clear-cut. Thus, criteria for 'similarity' are required, perhaps on a scale of increasing certainty. This issue was emphasized by comparisons rendered by *Smith* [1982] who matched metabolic profiles of antimalarial drugs from several species, including man. His results indicated that similarity was a function of numbers of matched peaks on mass spectrometric gas chromatograms, and that there was great similarity of profiles between nonhuman primates and humans, but poor comparability of profiles between rodents and humans.

A further difficulty stems from the fact that it is probably misleading to refer to humans as if we were just another strain of laboratory animals. It is known that great interindividual differences in metabolism occur in the

[1] It should be noted that data that appear discordant may not always be so. Apparent differences may be due to the nature of the experimental systems employed (e.g., different degrees of pathologic examination, or tests of different statistical power), and not to true biological differences.

human population. Referring to '*the* human data', when only a small sample of the human population has been examined, could lead to serious errors and add to the extent of uncertainty.

Because critical data are absent to make carefully reasoned choices of data from mouse, rat, canine, or other species to extrapolate qualitatively to humans, the selection still remains largely pragmatic. The most sensitive species and most serious endpoint may be appropriate as conservative societal value judgments about potential human hazard. It is probably wise always to recognize explicitly such selections as value judgments. However, the entire data base needs to be viewed with the perspective of the weight of evidence. This concept requires the proper consideration of no-observed-effect data in replicate studies and in other species. Certainly, replication of adverse effects with some modification of experimental design and with a manifest dose-response relationship would weigh more heavily in the selection of a surrogate species than contradictory data in another species.

*Quantitative Extrapolation and Thresholds*

Once qualitative data from laboratory animals have been interpreted to imply hazards to human health, there remains the estimation of the magnitude of the effect, if any, at doses generally much lower than those used experimentally. This is the phase referred to earlier as risk estimation. Toxicity studies require the use of relatively high doses in order to identify frank adverse effects with relatively few test organisms. Toxicity data are presented primarily in a dose-response context to obtain assurance that an effect is not spurious and to determine the relative hazard with changing dose. Classical toxicity studies are designed to find a no-observed-adverse-effect level (NOAEL) or operational threshold. The term 'threshold' has been used to imply many meanings, sometimes simultaneously. For example, there are thresholds of toxicity for individual organisms regardless of species; and there are population thresholds, which are levels of exposure below which no member of a population of defined size is likely to respond. A NOAEL from a rodent study is a good example of a population threshold. Difficulties arise, however, if one infers that the threshold for a small population of rodents is equivalent to a population threshold for a much larger population of humans. The population threshold in a group of individuals, whether rats or humans, is a function of the mechanisms of action of the toxicant, the distribution of susceptibilities in a genetically defined group of individuals, and the statistical and biological sensitivity of the assay. There are few substances for which the mechanism of intoxication is known in rodents and even fewer in humans. While the distribution of human sensitivities to a particular toxicant is unlikely to be known empirically, the genetic diversity of the human population leads one to anticipate wider ranges of response in humans than in laboratory rodents. In addition, a laboratory study of toxicity (and even a large epidemiology study) has a finite limit of detectability that is related to the number of subjects in each group. All of these factors argue strongly that it would be incautious to equate a toxicity threshold in a relatively small group of rats with a toxicity threshold in a large population of humans. This approach does not deny the existence of biological thresholds, but merely indicates our inability to identify them in the target population.

The classical interpretation of toxicity data would have equated the operational threshold of a rodent toxicity study to a population threshold in humans, and then would have called for the application of a safety factor expressed as an adjustment for the possible variations in the species extrapolation process and for the application of values related to the societal toleration of risks. Operational schemes for defining the safety of certain substances (e.g., food additives, pesticides) have been in existence for several decades. These schemes are based on the notion that imposition of certain minimum safety margins (e.g., 100-fold for a NOAEL derived from a chronic toxicity study in animals) leads to a level of intake that is 'acceptable' for humans. These schemes are said to provide 'safety', but no claim is made that they ensure absolute safety. Because they cannot ensure zero risk (no scheme could), these schemes are, in effect, risk assessment schemes that make no attempt to estimate residual risk. It is likely that the risks found 'acceptable' under these schemes may differ among various substances, depending on the nature of the toxic effects of the substances, the dose-response relationships for these effects, and the nature of the experimental methods used. The scheme thus probably provides little consistency in the treatment of different substances.

In recent years, use of safety factors or strict 'zero risk' goals has been replaced, at least for carcinogens, by mathematical extensions of observed dose-response curves. The biological theories supporting these dose-response models appear quite plausible; however, there still continues to be controversy about the various methods used to predict the shape of the dose-response curve at low doses. Presently the linearized multistage model [*Crump*, 1981] appears to have the most extensive biologic basis and the greatest number of supporters. As additional data become available about carcinogenesis, other models may become appropriate.

The theorized mechanisms of chemical carcinogenesis notwithstanding, the factors that render dose-response extrapolation scientifically justifiable should be considered equally applicable to all forms of toxicity for which the threshold for exposed humans remains unknown. The methods now used for carcinogenic agents provide many advantages over the safety assessment scheme, even though the numerical estimate of carcinogenic risk is uncertain. Perhaps most important is their use of dose-response data. This approach also ensures that the risk assessment activity is clearly distinguishable from the policy decisions concerning risk management. By contrast, use of specific safety margins such as 100 requires a policy choice, and the safety assessment scheme tends to disguise this fact and makes this decision appear scientific. An approach which is based on risk estimation would portray a more conceptually correct view in that chemical exposures would not be classified as 'safe' or 'unsafe', and it would also provide a basis for placing the magnitude of risks into broader societal contexts.

### *Judgmental Uncertainties in Risk Assessment*

At the outset, we mentioned that risk assessment was a synthesis of descriptions of hazards, evaluations of exposures, and estimations of risk. Its foundation is data held together and molded by concepts whose own data base is constantly evolving. Both the data and the concepts are sources of varying uncertainties which must be acknowledged and clearly stated to provide risk managers with balanced perspectives about risk conclusions.

Uncertainties are described by a number of expressions. Descriptions added to individual sets of data include a number of statistical terms such as probability values, upper confidence intervals, and best estimates. And in at least one case, statements about the quality of data bases have been described grossly by 'uncertainty factors' generally separated by one or more log units [NAS, 1977a, b]. What is lacking is a measure of the predictive strength of the toxicity tests across species: namely, the false-positive/false-negative statements. More severely lacking are presentation formats for a comprehensive array of risk conclusions with appropriate ranges of estimates, notations of severities, indications of the strength of the scientific support, and boundaries of uncertainties. Developments of such schemata may eventually prove as important to the proper utilization of risk assessment as the continued development of the biological concepts that serve as its bases and guideposts.

## *References*

Crump, K.: Statistical aspects of linear exptrapolation; in Richmond, Walsh, Copenhaven, Health risk analysis, pp. 381–392 (Franklin Institute Press, Philadelphia 1981).

National Research Council, Safe Drinking Water Committee: Drinking water and health, vol. 1, pp. 1–939 (National Academy of Sciences, Washington 1977a).

National Research Council: Principles and procedures for evaluating the toxicity of household substances. A Report of the committee for the Revision of NAS Publication 1138, pp. 1–130 (National Academy of Sciences, Washington 1977b).

Smith, C.C.: Extrapolation of drug metabolism data from animals to man; in Kuemmerle, Basic principles of clinical pharmacology (Ediozioni Medico Scientifiche, Torino 1982).

J.V. Rodricks, MD, Environ Corporation, Counsel in Health and Environmental Science, 1850 K Street, N.W., Suite 1050, Washington, DC 20006 (USA)

Safety Evaluation and Regulation of Chemicals. 1st Int. Conf., Boston, Mass., 1982, pp. 85–98 (Karger, Basel 1983)

# Difficulties in Extrapolating Toxicological Data from Animals to Man

*S. Garattini*

Istituto di Ricerche Farmacologiche 'Mario Negri', Milano, Italia

Animal toxicology studies collect data useful in predicting whether a chemical may be harmful when applied to man. Usually several animal species are used and the chemical is given at increasing doses for different lengths of time until toxic effects are observed. The minimal dose inducing an appreciable toxic effect in the most sensitive animal species is then compared with the maximal dose likely to be given to man. The ratio thus obtained is considered a safety margin: the higher the ratio the lower is the risk that the chemical is likely to induce toxicity in man.

This straightforward 'classic' approach is no longer tenable because comparing doses means comparing the amounts of a chemical outside the body. It is more logical instead to compare the amounts of the chemical present in the blood or target organs of animals and man. This involves comparing blood or tissue concentrations of a chemical inducing a noteworthy toxic effect in the most sensitive animal species with the concentration at the same site when the chemical is given at the maximal dose recommended in man.

This approach is dictated by the increasing knowledge showing that there may be no direct relation across animal species between the administered dose and the concentrations reached in various tissues. Chemicals are often absorbed, distributed, metabolized and excreted in a characteristic manner for each animal species. The same dose, given to different animal species and, in some cases, to different individuals of the same species, may result in different blood and tissue concentrations.

*Table I.* Safety ratios for MSG in plasma concentrations (AUC) assuming that the sensitivity of man to the neurotoxic effect of MSG is similar to that in newborn mice

| Animals species | Oral dose, mg/kg | AUC, μmol/ml × min | Safety ratio |
|---|---|---|---|
| Mouse | 500[a] | 318 | – |
| Man (prematures) | 1–11[b] | n.d. | very high |
| Man (infants) | 150[c] | 3.64 | 87 |
| Man (adults) | 30[d] | 1.69 | 188 |
| Man (adults) | 60[e] | 5.56 | 57 |
| | 120[f] | 8.00 | 40 |
| | 60[g] | 2.89 | 110 |
| | meal | 0.22 | 1,445 |

[a] $ED_{50}$ (dose producing hypothalamic lesions in 50% of the treated mice).
[b] These doses include the 99.9th percentile of MSG intake per meal for the population of the age 0–5 months [8].
[c] Calculated from the data of *Tung and Tung* [3].
[d] About 60 times the mean MSG intake per meal of the adult US population [8].
[e] About 10 times the 99th percentile MSG intake per meal of the adult US population [8].
[f] Unpalatable.
[g] Dissolved in tomato juice.

*Table II.* Kinetic parameters of plasma caffeine in mice and in man

| Caffeine 1 mg/kg p.o. | Plasma | | |
|---|---|---|---|
| | $C_{max}$, μg/ml | $T_{1/2}$, min | AUC, μg/ml × min |
| Man | 1.4 | 264 | 501 |
| Mouse | 1.1 | 56 | 81 |
| Ratio | 1.3 | 5 | 6 |

*Table III.* Saccharin concentrations in plasma and urine of rats and man

| Parameter | Rat[1] | Man[2] | Ratio |
|---|---|---|---|
| Administered dose, mg/kg | 5,000 | 2.10 | 2,385 |
| Plasma $C_{max}$, μg/ml | 54.5 | 0.35 | 156 |
| Plasma AUC, μg/ml × min | 863.1 | 2.24 | 385 |
| Urine $C_{max}$, μg/ml | 95,580 | 175 | 546 |
| Urine 24-hour excretion, mg/kg b.w. | 1,380 | 1.67 | 827 |

[1] Given in the diet at 5%.
[2] Given in solution with food (0.7 mg/kg × 3).

### *Examples of Relations between Doses and Concentrations*

The discrepancy between doses and concentrations across animal species with particular reference to man can be illustrated by a few examples.

High doses of monosodium glutamate (MSG) have neurotoxic effects in newborn mice [1]. 500 mg/kg given by gavage at a concentration of 10% to 7-day-old mice induce a lesion of the nucleus arcuatus of the hypothalamus in 50% of the treated animals. The dose required to obtain the same lesion in adult animals is considerably higher [2]. In man an oral dose of 120 mg/kg, although unpalatable, is apparently well tolerated although it represents ¼ of the toxic dose for infant mice. However, blood MSG levels reach a higher peak and persist longer in newborn mice than in both infant [3] and adult [4] humans. The reasons for this difference may be related to a limited capacity of the newborn mouse to metabolize MSG, and to the modality of administration [5].

As reported in table I the AUC (area under the curve) describing the blood levels of MSG in relation to time is several times greater in mice than in man compared to the administered dose of MSG. In contrast to mice [7], human infants metabolize MSG at a rate comparable to that of adults [3, 6]. Therefore, on the basis of doses the margin of safety would appear to be very low (about 3) but on the basis of blood concentrations it becomes quite high, over 50 for a dose that in man is 10 times the 99.9th percentile MSG intake per meal of the adult US population [8].

Caffeine, given at equal doses to mice and man (1 mg/kg orally), results in much higher blood concentrations in man than in mice. Table II indicates that the peak plasma level, the half-life and the area under the curve are several times greater in man than in mice after administration of the same dose.

Taking into account blood or tissue concentrations of some chemicals, the safety margin may become lower in relation to the doses. Saccharin has been shown to induce urinary bladder cancer when given in utero to rats at a concentration of 5% in the diet for the whole lifespan [9]. Table III shows the blood and urinary levels of saccharin achievable under this experimental condition. These can be compared with the blood and urine saccharin concentrations obtained in man during regular consumption [10]. The ratios indicate that the difference between blood or urine concentration in rats and man is less than might be expected on the basis of the difference between the doses.

### *Dose-Dependent Kinetics*

Extrapolating data from animals to man is not quite as simple as these considerations suggest because of a number of complications that may arise in relation to the chemical considered. With regard to kinetic parameters the high doses used in toxicology studies cannot be regarded as mere multiples of small doses. For several chemicals the kinetics are what is termed dose-dependent. The rise in blood or tissue levels is not proportional to the increase of the dose because there may be a limited capacity for metabolism or excretion.

*Table IV*. Brain area under the curve (AUC) for *d*- and *l*-fenfluramine (F) after administration of different doses of the racemic compound (the ratio between the considered dose and the lowest dose (5 mg/kg p.o.) is shown in parentheses)

| *d,l*-F, mg/kg p.o. | Brain AUC, μg/g × min | |
|---|---|---|
| | dF | lF |
| 5 | 587 | 53 |
| 13 (2.5) | 2,166 (3) | 513 (9) |
| 40 (8.0) | 22,661 (38) | 9,156 (172) |

An example is provided by fenfluramine, an anorectic drug which results in brain levels much higher than those expected on the basis of the increase in the dose [11]. As shown in table IV, this is particularly important for the *l*-isomer but is equally evident for the *d*-form. When the dose is increased 8 times the increased exposure (AUC) in the rat brain amounts to 38 and 172 times, respectively, for the *d*- and *l*-isomers of fenfluramine.

Caffeine also shows dose-dependent kinetics. In the oral dose range of 1–10 mg/kg, the rat is the only species showing this property [12]. When the dose of caffeine is increased by 5 and 10 times, the plasma AUC increases by about 17 and 45 times, respectively [13]. The effect may become even more marked in older animals; 1-year-old rats show a plasma AUC about 3 times higher than young rats given the same dose of caffeine (10 mg/kg orally) [14].

Extensive knowledge of the kinetics of chemicals may be helpful to calculate a 'kinetically equivalent dose' (KED). This may be defined as the dose which gives comparable plasma or tissue levels in different animal species. The KED of an anorectic dose of fenfluramine in man (40 mg or 0.7 mg/kg orally) is 1.2 and 3.2 mg/kg orally in rats on the basis of plasma peak level and AUC, respectively.

When kinetic data are available for different doses in several animal species, as with caffeine, it may be possible to establish a full range of KED. Table V shows the KED for caffeine in rats and man taking into account the plasma AUC. Because of the dose-dependent kinetics of caffeine in rats, the ratio between the KED for the two species changes as the dose rises. It is also clear that depending on the kinetic parameter selected the KED may not exactly overlap.

The plasma AUC is sometimes not the best indicator of 'dose-dependent' kinetics. For instance for VP16, a podophyllin derivative, there is a disproportionate rise in liver AUC as the dose rises, an effect which is not evident in other tissues [15] (table VI).

*Table V.* Man vs. doses (mg/kg) equivalence for caffeine (on the basis of AUC)

| Man | Rat |
|---|---|
| 0.22 | 1.84 |
| 1.00 | 4.82 |
| 1.60 | 6.50 |
| 2.50 | 8.63 |
| 5.00 | 13.42 |
| 7.50 | 17.37 |
| 10.00 | 20.86 |

*Table VI.* Area under the curve (AUC) and peak level of various tissues after administration of two doses of VP 16

| Tissue | VP 16 (13 mg/kg i.v.) | | VP 16 (40 mg/kg i.v.) | |
|---|---|---|---|---|
| | peak level μg/g | $AUC_{0-6h}$ μg/g × min | peak level μg/g | $AUC_{0-6h}$ μg/g × min |
| Liver | 41.9 ± 5.9 | 2,293 ± 307 | 212.0 ± 25.1 (5) | 28,464 ± 4,835 (12) |
| Kidney | 45.9 ± 11.7 | 3,633 ± 1,064 | 227.3 ± 21.1 (5) | 12,025 ± 2,111 (3) |
| Spleen | 11.0 ± 0.8 | 1,158 ± 143 | 41.2 ± 5.7 (4) | 2,841 ± 597 (2) |
| Heart | 29.4 ± 1.7 | 1,946 ± 323 | 99.1 ± 1.9 (3) | 3,818 ± 664 (2) |
| Brain | 3.0 ± 0.5 | 431 ± 46 | 5.4 ± 0.8 (2) | 404 ± 79 (1) |

The figures in parentheses express the factor of increase of the AUC obtained after the higher dose in respect to the lower dose. It is evident the disproportion of the increase in liver AUC in respect to other tissues.

Which kinetic parameter is most important for toxicological considerations? Obviously there can be no general answer since some toxicological effects may be related to the peak level while others may depend on the total exposure to the drug. The cytotoxicity of some anticancer agents depends on the concentration, but for other agents the duration of exposure is more important. Figure 1 illustrates one such case: VP16 is only slightly cytotoxic even at relatively high concentrations if the time of contact is short (24 h) but it becomes highly cytotoxic if the time of contact is prolonged (72 h) [16]. Such studies suggest that for the cytotoxic effect of VP16 the AUC may be a better indicator than the $C_{max}$.

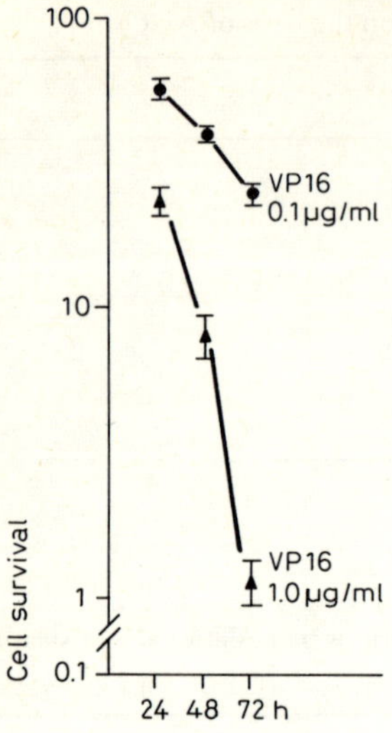


*Fig. 1.* Cytotoxicity of two concentrations of VP16 on Lewis lung carcinoma cells cultivated in vitro when different times of exposure have been utilized. It is evident that the time of exposure is more important than the concentration.

The ideal situation would be to know the concentration of the chemical under study at the receptor sites but this is precluded in most toxicological studies, particularly in man.

## *The Active Metabolites Dilemma*

For several chemicals it is not the administered chemical species that is responsible for the activity but one or more products formed from it in the body.

For example, benzopyrene per se is not a carcinogenic agent; it has to be hydroxylated to form the 'ultimate' carcinogenic species which interacts with DNA [17]. In several cases it may therefore be incorrect to compare the kinetics of a chemical in animals and man because what should be compared is the kinetics of a metabolite. The adriamycin derivative AD 32 (see fig. 2 for chemical structure) is cytotoxic not per se but because it is hydrolyzed to form the metabolite AD 41 (fig. 2). AD 41 is not only more cytotoxic than the parent compound but on account of its kinetic characteristics it remains in the body longer [18] (table VII).

Clobazam is a 1,5-benzodiazepine with a tranquillizing activity; it is N-demethylated in man to form N-desmethylclobazam, a metabolite with a longer half-life than the parent com-

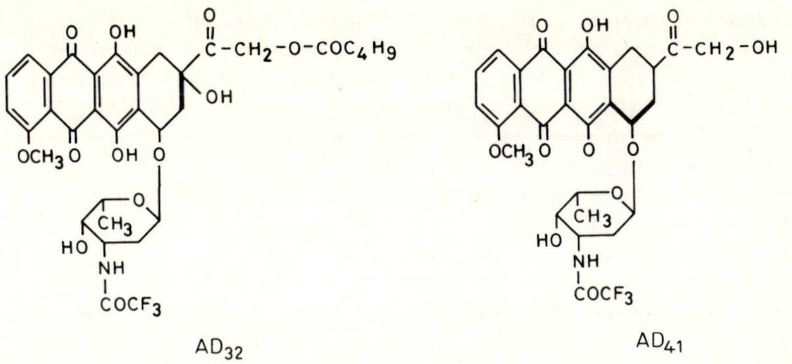

*Fig. 2.* Chemical structures of the adriamycin analogue AD 32 and its metabolite (AD 41) resulting from removal of the valeric acid.

*Table VII.* Kinetic parameters of AD 32 and AD 41 after administration of AD 32 (80 mg/kg i.v.) to mice bearing a 20-day-old Lewis lung carcinoma (3LL)

| 3LL | AD 32 | AD 41 |
|---|---|---|
| $C_{max}$, µg/g | 5.2 ± 0.6 (10 min) | 18.3 ± 1.5 (30 min) |
| AUC, µg/g × min | 161 ± 15 | 16,078 ± 4,197 |
| $T_{1/2}$, h | 0.34 | 5.82 |
| Cytotoxicity (µg/ml) | | |
| > 50% of inhibition of growth in vitro | 50 | 5 |

Cytotoxicity was studied in vitro on the same cancer cells. In vivo, AD 41 reaches concentrations in the tumor exceeding the cytotoxic level in vitro. The contrary is observed for AD 32. $C_{max}$ = Maximal concentration; AUC = area under the curve describing the relation between tumor drug level and time; $T_{1/2}$ = half-life.

pound which therefore accumulates during chronic treatment in man [19]. It would be incorrect to compare only clobazam levels because the levels of N-desmethylclobazam must be included in the evaluation. In this respect, it is relevant to underline that N-desmethylclobazam is not detectable in rats [20] but it is the major metabolite of clobazam in mice and guinea pigs [21]. The rat is thus clearly not a suitable species, at least from the kinetic point of view, from which to make predictions for the toxicity of clobazam in man. Caffeine is another chemical that is metabolized differently in different animal species. Table VIII compares the plasma AUC in man and mice, rats, rabbits and monkeys for caffeine and for the three dimethylxanthines, theophylline (1,3 DMX), theobromine (3,7 DMX) and paraxanthine (1,7 DMX) which are measurable in plasma [22]. Marked interspecies differences exist. None of these animal species can be taken as similar to man. This is underlined in table IX which gives the ratios between man and various animal species for caffeine and the three dimethylxanthines.

*Table VIII.* Plasma dimethylxanthines after caffeine (10 mg/kg oral)

| Animals species | Caffeine AUC | Theophylline AUC | Theobromine AUC | Paraxanthine AUC |
|---|---|---|---|---|
| Mouse | 1,022 | 36.12 | 479.4 | 204.9 |
| Rat | 2,098 | 192.6 | 272.4 | n.d. |
| Rabbit | 1,000 | 242.7 | 116.1 | 1,737.1 |
| Macaca | 4,645 | 6,487.5 | 747.3 | 417.0 |
| Man (5 mg/kg) | 4,706 | 300.8 | 424.3 | 1,757.5 |

*Table IX.* Ratios between human plasma AUC of caffeine (1,3,7 TMX), theophylline (1,3 DMX), theobromine (3,7 DMX) and paraxanthine (1,7 DMX) and plasma AUC of the considered animal species: it is evident that no animal species resemble man as far as caffeine metabolism is concerned

| Caffeine 10 mg/kg p.o. | Ratios in respect to man | | | |
|---|---|---|---|---|
| | 1,3,7 TMX | 1,3 DMX | 3,7 DMX | 1,7 DMX |
| Mouse | 9.2 | 16 | 1.6 | 17 |
| Rat | 4.4 | 3 | 3.0 | – |
| Rabbit | 9.4 | 2.4 | 7.2 | 2 |
| Macaca | 2.0 | 0.09 | 1.1 | 8 |
| Plasma AUC | | | | |

*Table X.* Plasma peak value and AUC of diphenylhydantoin (DPH) and its monohydroxylated metabolite (HPPH) in rats and in man

| Species | Oral dose mg/kg | Peak value, μg/ml | | AUC, μg/ml × h | |
|---|---|---|---|---|---|
| | | DPH | HPPH | DPH | HPPH |
| Rat | 100 | 27.3 | 2.1 | 324.7 | 26.6 |
| Man | 6 | 6.2 | 0.9 | 174.5 | 35.2 |

Urinary metabolites present more complexities. For instance 1,3,7-DAU (1,3,7 trimethyl, 5 formylamino, 6 aminouracil), a caffeine metabolite with mild cytotoxic effects [23] is excreted in large quantities by rats but not by man [24]. In contrast, man excretes relatively high concentrations of 5-acetylamino,6-amino,1-methyluracil [25]. It is likely that some of the most severe

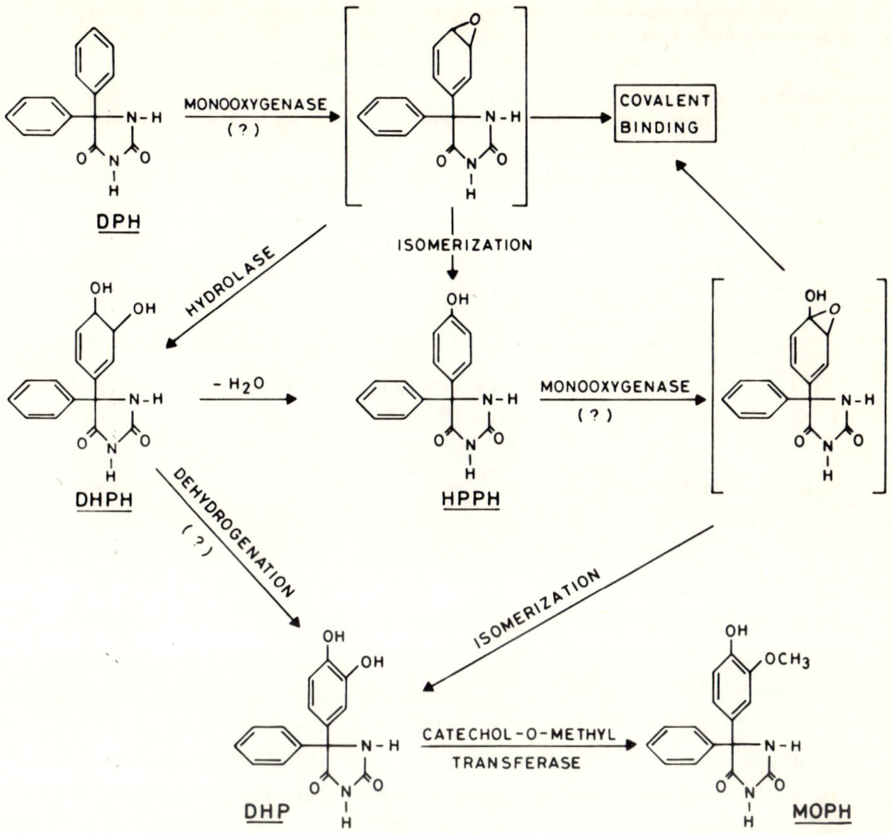


*Fig. 3.* Known and postulated diphenylhydantoin (DPH) in vivo metabolism through formation of chemically reactive intermediates. HPPH = Monohydroxylated metabolite; DHP = dihydroxylated metabolite.

toxic effects are due not to stable chemical species but to those which are relatively unstable and therefore have the potential to reach and covalently bind to cellular constituents, particularly macromolecules (proteins and nucleic acids).

Diphenylhydantoin (DPH) is an antiepileptic agent which is suspected of inducing teratogenic effects in man. The metabolism of DPH involves (fig. 3) a hydroxylation step, among others, present in rats and man. Under certain experimental conditions a certain parallelism can be seen between the mutagenicity [26] and covalent binding to macromolecules [27] of DPH (and its hydroxylated metabolite HPPH). As shown in table X, the plasma AUC

*Table XI.* Plasma kinetic parameters of adriamycin (ADM) in cancer patients treated i.v. with different doses of ADM

| ADM dose range, mg/m$^2$ | $C_o$, µg/ml | $\beta T_{1/2}$, min | AUC, µg/ml × min |
|---|---|---|---|
| 20–30 | 2.3–9.3 | 195–293 | 58–214 |
| 35–46 | 1.4–11.7 | 95–267 | 63–370 |
| 48–60 | 6.3–15.8 | 81–223 | 105–235 |

$C_o$ = Concentration extrapolated at 0 time $\beta T_{1/2}$ = half-life; AUC = area under the curve.

of DPH and HPPH of therapeutic doses in man are of the same order of magnitude as those obtainable in rats with 100 mg/kg [28].

*Individual Variability*

In some cases the individual variability is limited but in others it may reach several orders of magnitude. Table XI shows the variability of kinetic parameters of adriamycin in man under realistic conditions of use [29]. From 20 to 60 mg/m$^2$ the kinetic parameters almost overlap, indicating that the $C_{max}$ and the AUC may be higher in a patient given the lowest dose than in one treated with the highest dose. It may be difficult to define the optimal therapeutic dose or the maximal tolerated dose of this drug for an individual patient unless blood levels of adriamycin are carefully monitored.

*Equal Sensitivity?*

'If it were possible to determine kinetic parameters of a chemical in animals and man, are equal concentrations of the substance (and/or its active metabolites) equivalent to equal toxicological effects?' Do similar concentrations of chemicals elicit similar biological effects in various animal species? No universal reply can be given because the sensitivity to chemicals may sometimes be similar but in other cases it is different when man is compared to other animal species or even when different individuals of the same animal species are compared.

Saccharin given at high doses in the diets (1–5%) induces marked immunodepression in rats [30]. Similarly, saccharin prevents the mitogenic activity of PHA in mouse or rat lymphocytes [31]. However, as shown in figure 4, the concentrations of saccharin effective in rats do not have the same effects on human lymphocytes. Thus, equal concentrations do not mean equal

*Table XII.* Inhibition of 9 human ovarian carcinomas by VP 16 at therapeutic concentrations (the inhibition of thymidine ($^3$H-TdR) uptake which is not significant is reported in parentheses)

| Human ovarian carcinoma | % inhibition $^3$H-TdR uptake[1] |
|---|---|
| 1 | 42 |
| 2 | (3) |
| 3 | 28 |
| 4 | (0) |
| 5 | (20) |
| 6 | 66 |
| 7 | 53 |
| 8 | 30 |
| 9 | (11) |

[1] After 2.8 µg/ml VP 16 (plasma AUC).

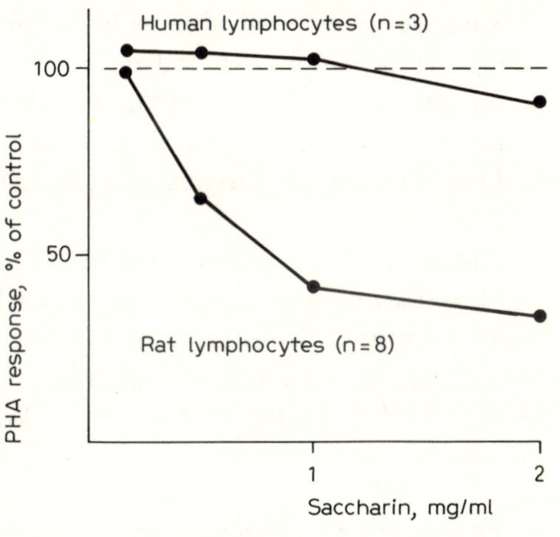


*Fig. 4.* Responsiveness to phytohemagglutinin (PHA) of human and rat lymphocytes exposed to similar concentrations of saccharin.

toxic effects. These data also suggest that the lack of activity in an animal species does not always imply a lack of activity in man. Sometimes cells from different individuals of the same species show different degrees of sensitivity to the cytotoxicity of a chemical. Ovarian carcinoma cells obtained from different patients are killed in different percentages by VP16 at concentrations which are achieved at therapeutic doses in man (table XII). Consequently, both the therapeutic and the toxic outcome may vary widely in different patients at the same plasma or tumor concentrations of VP16.

*Table XIII.* Similarity of warfarin dose, concentration and anticoagulant activity in rats and man

| | Rat | Man |
|---|---|---|
| Warfarin dosage, mg/kg | 0.8–1.0 | 1.0–1.5 |
| Warfarin level, μg/ml blood | 1.5–3.0 | 5.0–7.5 |
| Prothrombin complex activity, % | 10–30 | 10–20 |

Despite these limitations there are agents for the activity of which can be predicted across animal species. Table XIII shows that similar doses of warfarin induce similar blood levels and that similar concentrations result in comparable anticoagulant activity [32, 33].

## *Conclusions*

By now, it is clear that the toxic doses of a chemical must be integrated by data indicating the corresponding concentrations in blood or tissues across animal species. Thus, a considerable amount of kinetic and metabolic data should be collected for a better interpretation of the toxicological findings. This is now possible on account of the considerable progress in modern analytical techniques.

Data cannot be extrapolated in general terms, but must be obtained separately for each chemical. Extrapolation is a complex process which requires the contribution of several disciplines. Routine work in toxicology is meaningless if it is not accompanied by adequate supporting studies. Similarly, the use of general mathematical models is not justified when no knowledge is available on species differences in drug kinetics, metabolism, sensitivity and mechanism of action.

There is a need to develop and to test new methodologies, such as the approach proposed in this paper, particularly when chronic effects are considered. Toxicologists should not give the public a false feeling of security or cause undue alarm, when in fact only vague trends and hypotheses are available.

## *References*

1 Olney, J.W.: Excitotoxic amino acids. Research applications and safety implications; in Filer, Garattini, Kare, Reynolds, Wurtman, Glutamic acid. Advances in biochemistry and physiology, pp. 287–319 (Raven Press, New York 1979).

2 Takasaki, Y.; Matsuzawa, Y.; Iwata, S.; O'Hara, Y.; Yonetani, S.; Ichimura, M.: Toxicological studies of monosodium l-glutamate in rodents: Relationship between routes of administration and neurotoxicity; in Filer, Garattini, Kare, Reynolds, Wurtman, Glutamic acid. Advances in biochemistry and physiology, pp. 255–275 (Raven Press, New York 1979).

3 Tung, T.-C.; Tung, K.S.: Serum free amino acid levels after oral glutamate intake in infant and adult humans. Nutr. Rep. Int. *22* (1980).

4 Airoldi, L.; Bizzi, A.; Salmona, M.; Garattini, S.: Attempts to establish the safety margin for neurotoxicity of monosodium glutamate; in Filer, Garattini, Kare, Reynolds, Wurtman, Glutamic acid. Advances in biochemistry and physiology, pp. 321–331 (Raven Press, New York 1979).

5 Garattini, S.: Evaluation of the neurotoxic effects of glutamic acid; in Wurtman, Wurtman, Nutrition and the brain, vol. 4, pp. 79–124 (Raven Press, New York 1979).

6 Airoldi, L.; Bonfanti, M.; Ghezzi, P.; Salmona, M.; Garattini, S.: Effect of oral monosodium glutamate on glutamic acid levels in the nucleus arcuatus of the hypothalamus and on serum osmolality of adult and infant mice. Toxicol. Lett. *7:* 107–111 (1980).

7 Filer, L.J., Jr.; Baker, G.L.; Stegink, L.D.: Metabolism of free glutamate in clinical products fed infants; in Filer, Garattini, Kare, Reynolds, Wurtman, Glutamic acid. Advances in biochemistry and physiology, pp. 353–362 (Raven Press, New York 1979).

8 Committee on GRAS List Survey – Phase III: Application of a probabilistic method in estimating daily intakes of certain GRAS substances. Food and Nutrition Board, Division of Biological Sciences, Assembly of Life Sciences, National Research Council, National Academy of Sciences, Washington, D.C. (in manuscript dated August, 1976).

9 Arnold, D.L.; Moodie, C.A.; Grice, H.C.; Charbonneau, S.M.; Stavric, B.; Munro, I.C.: Long term toxicity of orthotoluenesulfonamide and saccharin in the rat. Toxicology forum (The Eppley Institute for Research in Cancer, Omaha 1977).

10 Pantarotto, C.; Salmona, M.; Garattini, S.: Plasma kinetics and urinary elimination of saccharin in man. Tox. Lett. *9:* 367–371 (1981).

11 Caccia, S.; Dagnino, G.; Garattini, S.; Guiso, G.; Madonna, R.; Zanini, M.G.: Kinetics of fenfluramine isomers in the rat. Eur. J. Drug Metab. Pharmacokinet. *6:* 297–301 (1981).

12 Garattini, S.: Concluding remarks. Extrapolation of toxicological data from animals to man; in Bartosek, Guaitani, Pacei, Animal in toxicological research, pp. 201–208 (Raven Press, New York, 1982).

13 Latini, R.; Bonati, M.; Castelli, D.; Garattini, S.: Dose-dependent kinetics of caffeine in rats. Toxicol. Lett. *2:* 267–270 (1978).

14 Latini, R.; Bonati, M.; Marzi, E.; Tacconi, M.T.; Sadurska, B.; Bizzi, A.: Caffeine disposition and effects in young and one year old rats. J. Pharm. Pharmac. *32:* 596–599 (1980).

15 Colombo, T.; Broggini, M.; Torti, L.; Erba, E.; D'Incalci, M.: Pharmacokinetics of VP16-213 in Lewis lung carcinoma bearing mice. Cancer Chemother. Pharmac. *7:* 127–131 (1982).

16 D'Incalci, M.; Erba, E.; Vaghi, M.; Morasca, L.: In vitro cytotoxicity of VP16 on primary tumor and metastasis of Lewis lung carcinoma. Eur. J. Cancer Clin. Oncol. *18:* 377–380 (1982).

17 Pelkonen, O.; Boobis, A.R.; Nebert, D.W.: Genetic differences in the binding of reactive carcinogenic metabolites to DNA; in Jones, Freudenthal, Carcinogenesis, vol. 3. Polynuclear aromatic hydrocarbons, pp. 383–400 (Raven Press, New York 1978).

18 Abbruzzi, R.; Rizzardini, M.; Benigni, A.; Barbieri, B.; Donelli, M.G.; Salmona, M.: Possible relevance of N-trifluoroacetyladriamycin (AD 41) in the antitumoral activity of N-trifluoroacetyladriamycin-14-valerate (AD 32) in tumor bearing mice. I. Pharmacokinetic evidence. Cancer Treat. Rep. *64:* 873–878 (1980).

19 Rupp, W.; Badian, M.; Christ, O.; Hajdu, P.; Kulkarni, R.D.; Taeuber, K.; Uihlein, M.; Bender, R.; Vanderbeke, O.: Pharmacokinetics of single and multiple doses of clobazam in humans. Br. J. clin. Pharmac. *7:* 51S–57S (1979).

20 Caccia, S.; Guiso, G.; Garattini, S.: Brain concentrations of clobazam and N-desmethylclobazam and antileptazol. J. Pharm. Pharmac. *32:* 295–296 (1980).

21 Ballabio, M.; Caccia, S.; Garattini, S.; Guiso, G.; Zanini, M.G.: Antileptazol activity and kinetics of clobazam and N-desmethyl-clobazam in the guinea-pig. Archs int. Pharmacodyn. Thér. *253:* 192–199 (1981).

22 Bonati, M.; Latini, R.; Galletti, F.; Young, J.F.; Tognoni, G.; Garattini, S.: Caffeine disposition after oral doses. Clin. Pharmacol. Ther. *32:* 98–106 (1982).

23 Garattini, S.; Erba, E.; Morasca, L.; Peri, G.; Mantovani, A.; Filippeschi, S.; Spreafico, F.; Arnaud, M.: In vitro and in vivo cytotoxicity of 6-amino-5-formylmethylamino-1,3-dimethyl uracil, a uracilic metabolite of caffeine. Toxicol. Lett. *10:* 313–319 (1982).

24 Latini, R.; Bonati, M.; Marzi, E.; Garattini, S.: Urinary excretion of an uracilic metabolite from caffeine by rat, monkey and man. Toxic. Lett. *7:* 267–272 (1981).

25 Callahan, M.M.; Robertson, R.S.; Lavin, M.J.; Yesair, D.W.: Human metabolism of caffeine; in Federation of American Societies for Experimental Biology, Abstracts of Papers, 63rd Annu. Meeting, Dallas, Texas 1979, abstract 1877.

26 Sezzano, P.; Raimondi, A.; Arboix, M.; Pantarotto, C.: Mutagenicity of diphenylhydantoin and some of its metabolites towards *Salmonella typhimurium* strains. Mutation Res. *103:* 219–228 (1982).

27 Pantarotto, C.; Arboix, M.; Sezzano, P.; Abbruzzi, R.: Studies on 5,5-diphenylhydantoin irreversible binding to rat liver microsomal proteins. Biochem. Pharmac. *31:* 1501–1507 (1982).

28 Arboix, M.; Pantarotto, C.: Unpubl. data.

29 Piazza, E.; Donelli, M.G.; Broggini, M.; Sessa, C.; Natale, N.; Ottolenghi, L.; Marsoni, S.; Libretti, A.; Mangioni, C.; Morasca, L.: Early phase pharmacokinetics of doxorubicin (adriamycin) in plasma of cancer patients during single- or multiple-drug therapy. Cancer Treat. Rep. *64:* 845–854 (1980).

30 Luini, W.; Mantovani, A.; Garattini, S.: Effects of saccharin on primary humoral antibody production in rats. Toxicol. Lett. *8:* 1–6 (1981).

31 Mantovani, A.; Luini, W.; Candiani, G.P.; Salmona, M.; Spreafico, F.; Garattini, S.: In vitro effects of saccharin on cell-mediated host defence mechanisms. Toxicol. Lett. *5:* 287–295 (1980).

32 Breckenridge, A.: Oral anticoagulants – the totem and the taboo. Br. med. J. *i:* 419–423 (1976).

33 Yacobi, A.; Wingard, L.B., Jr.; Levy, G.: Comparative pharmacokinetics of coumarin anticoagulants. X. Relationship between distribution, elimination and anticoagulant action of warfarin. J. pharm. Sci. *63:* 868–872 (1974).


Dr. S. Garattini, Istituto di Ricerche Farmacologiche 'Mario Negri', Via Eritrea 62, I-20157 Milan (Italy)

Safety Evaluation and Regulation of Chemicals. 1st Int. Conf., Boston, Mass., 1982, pp. 99–105 (Karger, Basel 1983)

# Perceived Risk and Technological Hazards

## A Review Essay on the Literature

*Vincent T. Covello*[1]
National Science Foundation, Washington, D.C., USA

A truly unexpected result came out of the Kemeny Commission's study of Three Mile Island. A group that set out to investigate a technology ended up talking about people... In the commission's own words, 'It became clear that the fundamental problems were people-related problems.'

Editorial, The Washington Post, October 31, 1979

Behavioral and social scientists in several countries are currently grappling with several people-related questions related to risk: What factors influence individual perceptions of risk? What accounts for anomalies in the way individuals and groups behave when faced with ostensibly comparable risks – such as the risks of nuclear power, dam failures, or earthquakes? What weight should decision-makers attach to public perceptions of risk in determining how safe is safe enough? Are there ways to increase our capacity for dealing with technological risks in a rational manner? What follows is a review of the behavioral and social science literature pertaining to these questions [see *Covello and Abernathy,* in press].

### *Human Intellectual Limitations*

Research suggests that people do not cope well when confronted with risk problems and decisions. Intellectual limitations and the need to reduce

[1] This review draws heavily on the work of *Paul Slovic, Baruch Fischhoff,* and *Sarah Lichtenstein,* and I would like to acknowledge this contribution. I would also like to thank *Jeryl Mumpower, Mark Abernathy, Joshua Menkes,* and *Jiri Nehnevajsa* for their help. The views expressed in this paper are exclusively my own and do not necessarily represent the views of the National Science Foundation.

anxiety often lead to the denial that risk and uncertainty exist and to unrealistic oversimplifications of essentially complex problems. To simplify risk problems, people use a number of inferential or judgmental rules, known technically as heuristics. Two of the most important are (1) information availability, or the tendency for people to judge an event more frequent if instances of it are easy to imagine or recall, and (2) representativeness, or the tendency of people to assume that roughly similar activities and events (such as nuclear power technologies and nuclear war) have the same characteristics and risks. These judgmental operations enable people to reduce difficult probabilistic and assessment tasks to simpler tasks; however, these judgmental operations also lead to severe and systematic biases and errors.

One bias associated with information availability is that people have difficulty imagining low probability/high consequence events happening to themselves. Unless people have been made graphically aware of the risks, typically through past experience, they are unlikely to take protective action. A classic example is the observed reluctance of flood plain residents to purchase low cost flood insurance. Compounding the problem is the difficulty people have understanding and interpreting probabilistic information. People residing in 100-year flood plains, for example, typically believe that a recent severe flood precludes the possibility of another severe flood in the near future. According to folk wisdom, but not according to probability theory, lightning never strikes the same place twice.

Biases associated with information availability have also been used to explain the results of studies in which people were asked to judge the frequency of various causes of death, such as accidents, tornadoes, and diseases. These studies show that the risks of low frequency events tend to be overestimated, and that the risks of high frequency events are underestimated. People underestimate fatalities caused by asthma, stroke, and diabetes, and overestimate fatalities from homicides, botulism, fires, snake bites, tornadoes, and abortion. Overestimated causes of death tend to be dramatic, sensational, and of interest to the media, whereas underestimated causes of death tend to be unspectacular, undramatic, and of little interest to the media.

Researchers have also pointed out that information availability biases may cause risk information campaigns and educational efforts to work at cross-purposes. Information may heighten the imaginability and consequently the perceived probability of a rare event, even when the information is designed to assure individuals that the event is unlikely. A package insert listing all the risks of taking a drug, or a published report describing safety precautions taken at a DNA research laboratory, may only serve to increase concern about the substance or activity. By identifying previously unknown ways in which things can go wrong, the information provider takes the chance that people will incorrectly assess the information, i.e., that they will consider the event more likely as a result of increased knowledge. Unfortunately, few researchers have critically examined the controversial hypothesis implicit in this work – that the very discussion of a low-probability hazard increases the judged probability of the hazard, regardless of what the evidence indicates.

Disputes and controversies about risk are made all the more difficult by another psychological mechanism: once beliefs are formed, individuals will frequently structure and distort the interpretation of new evidence, and will often resist disconfirming information. People tend to dismiss evidence contradicting their beliefs as unreliable, erroneous, and unrepresentative. The accident at Three Mile Island, for example, provided confirming evidence for those already convinced that nuclear power technology is safe. The accident also reinforced the beliefs of those who believed that nuclear power technology is dangerous. Convincing people that a hazard they fear is safe is extremely difficult even under the best conditions. Any accident or mishap, no matter how small, is seen as proof of high risk.

### *Overconfidence*

A second set of risk perception findings addresses the problem of overconfidence. Researchers have shown that experts and laypersons are typically overconfident about their risk estimates. In one study, participants were asked to state the odds that they were correct in judging which of two lethal events was more frequent. Most people claimed that the odds of their being wrong were 100:1 or greater. In actuality, people were wrong about one out of every eight times. Such overconfidence can produce serious judgmental errors, including judgments about how much is known about the hazard and about how much needs to be known. Of equal or greater importance, overconfidence leads people to believe that they are comparatively immune to common hazards. Studies show (1) that most people rate themselves among the most skillful and safe drivers in the population; (2) that people rate their own personal risk from several common household hazards as lower than the risk for others in society; (3) that people judge themselves average or above average in their ability to avoid bicycle and power lawnmower accidents, and

(4) that people underestimate and are extremely unrealistic about their chances of having a heart attack. In general, people underestimate the risks of activities that they perceive to be familiar and under their personal control, such as automobile driving.

Overconfidence has also been used to explain in part the observed reluctance of about 80–90% of the US driving population to wear seat belts. Unfortunately, few empirical studies have examined this issue or the more general relationship between perceived risk and protective behavior. A new study by *Slovic, Lichtenstein, and Fischhoff* of seat belt usage is, however, exploring and empirically testing the hypothesis that seat belt usage will increase if the public is presented with information about the lifetime risks of driving instead of information about the risks of taking a single trip.

### *Expert and Non-Expert Estimates of Risk*

A third set of findings bears on expert and non-expert estimates of risk. A consistent result is that technical experts and non-experts differ substantially in their risk estimates. Risk estimates of technical experts are closely correlated with annual fatality rates, whereas the risk estimates of non-experts are only moderately to poorly correlated with annual fatality rates. In explaining these differences, researchers have identified several factors other than annual fatality rates that influence public perceptions of risk. Risks are perceived to be higher if the activity is perceived to be involuntary, catastrophic, not personally controllable, inequitable in the distribution of its risks and benefits, unfamiliar, and highly complex. Other factors influencing risk perceptions are whether the adverse effects are immediate or delayed, whether exposure to the hazard is continuous or occasional, whether the technology is perceived to be necessary or a luxury, whether the adverse effects are well-known or uncertain, and whether the activity is certain to be fatal.

Several studies have shown that these dimensions of risk are closely related to one another. Such correlations have prompted several research groups to reduce the various dimensions of risk to a smaller number of factors. One study identified at least two factors: the level of technological complexity and the hazard's severity or catastrophic potential. In a follow-up study that examined a larger set of hazards and risk characteristics, *Slovic, Fischhoff, and Lichtenstein* found three factors: familiarity, dread, and the number of people exposed to the hazard. In an on-going European study of risk perception, *Vlek and Stallen* identified several additional factors influencing risk perception and risk acceptability, including the beneficiality of the technology and the degree to which protection is provided by institutional means. In spite of these different findings, it is clear that a hazard's catastrophic potential is uppermost in the minds of people. Since catastrophic events may threaten the survival of individuals, families, societies, and the species as a whole, such concern may be quite justifiable.

Analysis of inter-correlations between the various dimensions of risk have also led researchers to challenge Starr's well-known proposition that the risks of voluntary activities are more acceptable to the general public than the risks of involuntary activities. One problem with this proposition is that voluntary risks are also perceived by the public to be controllable, equitable, familiar, and noncatastrophic. These correlations suggest in turn that the observed greater willingness of the public to accept voluntary risks may be due to these other factors and not to the voluntary nature of the activity.

### *Risk Perception and Nuclear Power: A Case Study*

To date, most of the research on risk perception has focused on nuclear power. These studies have produced several important findings. First, researchers have shown that nuclear power has nearly all the characteristics associated with highly perceived risk. The risks of nuclear power are perceived to be involuntary, delayed in their consequences, unknown, uncontrollable, unfamiliar, potentially catastrophic, inequitable, and certain to be fatal. Public perceptions of nuclear power contrast sharply with non-nuclear sources of electric power, which are perceived to be noncatastrophic, familiar, controllable, and comparatively safe.

Second, researchers have shown that disputes about nuclear power are often about values and goals that far transcend issues of health and safety. Many people are concerned about nuclear power not because of its specific risks but because of its association with nuclear weapons, with highly centralized political and economic systems, and with technological elitism. The debate about nuclear power is also colored by social class – people with lower socioeconomic status are less supportive of nuclear power than those with higher socioeconomic status; by sex – women are less supportive of nuclear power than men, and by concerns about the credibility of institutions charged with estimating, evaluating, and managing the risks.

Despite these concerns, research studies consistently show that the public, by a margin of 2 and sometimes 3 to 1, supports nuclear power, even in the aftermath of Three Mile Island. Somewhat counterintuitively, researchers have also found that people living within the vicinity of a nuclear power plant (and therefore presumably subject to the greatest objective risk) are more supportive of nuclear power than those living farther away. In explaining this finding it has been proposed that people living near power plants receive greater economic benefits, that they experience greater cognitive dissonance, and that they have had their worst fears assuaged by a history of accident-free operations. Interestingly, those who are least supportive of nuclear power live in areas where power plants are under construction or being planned. One policy implication arises from these findings. In several countries, including France, proposals are currently being considered to compensate those who live in the vicinity of nuclear power plants. If the intention is to win wider public acceptance, then the policy is misdirected. Those living nearest to the power plant are already supportive and little would be gained by compensating them. By comparison, compensating those who are least supportive – i.e., those living in areas where power plants are under construction or being planned – might have a major impact. Such a policy could, of course, also backfire by providing support for the belief that the risks of nuclear power are indeed substantial.

### *Conclusion*

One conclusion that can be drawn from this literature review is that conflicts about technological risks appear to be partially rooted in the different risk analysis methods and approaches used by technical experts and nonexperts. Technical experts often implicitly and sometimes explicitly assign equal weight to hazards that take many lives at one time and to hazards that take many lives one at a time; nonexperts typically assign greater weight to hazards that take many lives at one time, i.e., to catastrophes. Technical experts often implicitly and sometimes explicitly assign equal weight to statistical and known deaths; nonexperts typically assign greater weight to known deaths. It is interesting in this regard to note the high levels of public concern and massive allocations of resources devoted to rescuing an identifiable person lost at sea. Technical experts often implicitly and sometimes explicitly assign equal weight to voluntary and involuntary risks; nonexperts typically assign greater weight to involuntary risks. Technical experts typically express risks in quantitative terms, and use computational and experimental methods to identify, estimate, and evaluate the risks; nonexperts typically express risks in qualitative terms, and use intuitive and impressionistic methods to identify, estimate, and evaluate the risks. Technical experts typically believe that quantitative estimates of risk should be the prime consideration in risk acceptability decisions; nonexperts typically believe that quantitative estimates of risk should be only one among several quantitative and qualitative considerations in risk acceptability decisions. Technical experts often implicitly and sometimes explicitly assign the same weight to different ways of dying; nonexperts typically feel that some ways of dying are worse than others. How one dies, and with how much suffering, is as important as where and when. Public risk acceptance and the success of risk management policies are likely to hinge on whether these differences are taken into account in the decision-making process.

## *Reference*

Covello, V.; Abernathy, M.: Risk analysis and technological hazards. A policy related bibliography; in Whipple et al., Technological risk assessment (Sijthoff & Noordhoff, Winchester, 1983).

V.T. Covello, MD, PhD, National Science Foundation, 1800 G. Street NW, Washington, DC 20006 (USA)

Safety Evaluation and Regulation of Chemicals. 1st Int. Conf., Boston, Mass., 1982, pp. 106–111 (Karger, Basel 1983)

# Priority Setting for Testing Chemicals

*Gib Bogle, Talbot Page*

Environmental Quality Laboratory, California Institute of Technology, Pasadena, Calif., USA

This paper addresses the problem of prioritizing chemicals for testing. To be concrete, consider the problem faced by the National Toxicology Program. This program has a budget of about $60 million to test chemicals for toxic effects. There is a universe of about 70,000 commercial chemicals to choose from, and this universe might be expanded to include natural chemicals and inadvertent by-products. There are dozens of effects which might be tested for, including carcinogenesis, mutagenesis, teratogenesis, fetotoxicity, reproductive failure, nervous system disorder, and pulmonary disorder. And, finally, there are hundreds of tests and test systems, ranging in cost from about $100 to $750,000.

Loosely stated, the objective of prioritization is to choose chemicals to be tested and tests to be performed in a way to maximize the value of information obtained from testing, subject to the budget constraint. In our approach to this problem we combine ideas from decision theory and statistical estimation theory. From decision theory we take a decision tree approach, with judgemental probabilities and sequential analysis. From statistical estimation theory, we take our criterion – a criterion analogous to minimizing an estimated variance.

The approach differs from others to priority setting in several ways. Our approach is to optimize the choice or design of the priority process (or to choose the best decision tree among a large number of possible decision trees). In *Weinstein* [5], the optimization is over a decision whether to test a chemical for a given priority process. In Weinstein's approach, which is probably the closest to ours, the criterion is to minimize the expected cost of final actions. This means that expected regulatory actions and inactions, along with the costs of chemical benefits foregone, must be estimated as a part of the priority process. By using a criterion of misclassification cost, the costs of final actions, which are expensive to estimate for many chemicals and highly uncertain at the time of prioritization, play a minor and indirect role. The justification for our approach is that gathering more than rudimentary information on the costs and benefits of final (regulatory) actions for 70,000 chemicals at the first stage of prioritization would not be efficient in terms of the expected value of this information.

To our knowledge, other existing approaches to prioritization for testing [1–4, 6] do not attempt to optimize.

*Definition*

A priority process can be summarized as a decision tree, as follows:

| | |
|---|---|
| $S = \{s_1, s_2, \ldots, s_N\}$ | Set of N categories of exposure and toxicity to which a chemical may belong. |
| $T = \{t_1, t_2, \ldots, t_K\}$ | Set of tests.[1] (This set includes the null test, which costs nothing and always has the same outcome.) |
| $R_k = \{r_{k1}, r_{k2}, \ldots, r_{k,m(k)}\}$ | Set of m(k) possible results of test k. |

An L stage priority process $\pi$ is defined by specifying a test for stage 1; a test at stage l for each possible sequence of outcomes from the preceding l–1 stages; and a recommended test for each outcome of the L stage priority process.

*Priority Processes Viewed as Classification Devices*

A test k is characterized by an m(k) by N array of probabilities, called the performance characteristics:

| | |
|---|---|
| $P(r_{ki} \mid s_j) =$ | probability of test k having outcome $r_{ki}$ when the true category of a chemical tested is $s_j$. |

The current state of knowledge about any universe (or subuniverse) of chemicals is represented by a probability distribution over the N chemical types or categories:

| | |
|---|---|
| $P(s_j) =$ | probability that a chemical drawn randomly from the universe of chemicals will fall into category $s_j$. $P(s_j)$ is the prevalence of exposure and toxicity of type j. |

[1] A test may be a battery of tests, a data element, or a battery of data elements.

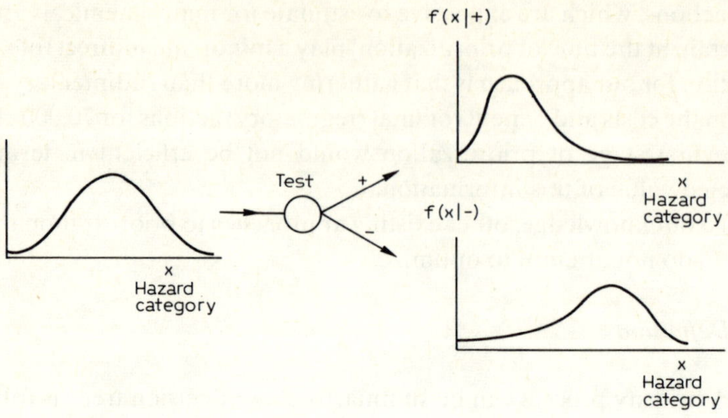


*Fig. 1.* A simple test.

The probability that a chemical chosen at random from the initial collection (the universe) being put into the $i^{th}$ result category by the $k^{th}$ test is

$$\sum_{j=1}^{N} P(r_{ki}\;\; s_j)\, P(s_j).$$

A test k transforms a prior distribution $P(s_j)$ into a set of posterior distributions $P(s_j \mid r_{ki})$, $i=1, \ldots m(k)$.

The increase in our level of information about the chemicals is manifested in the increased localization (decreased standard deviation) of each of the posterior distributions in comparison to the prior distribution. As a diagrammatic illustration, consider a continuous distribution f(x) and a test with two possible results, as shown in figure 1. The sum of the two posterior distributions, weighted by the total probabilities of each result, is the prior distribution.

As long as tests yield information, repeated tests will produce a proliferation of outcome categories of chemicals with ever-sharpening probability distributions. Eventually, if each test conveys information, each of the multitude of distributions would have zero standard deviation, in other words the true type of every chemical would be known with certainty. The existence of testing costs and a budget constraint means that such a perfect sorting cannot be achieved. The aim of a prioritization and testing program is to sort, as well as possible for a given budget, the universe of chemicals under consideration into the N types of true classification.

*Comparability and Misclassification Cost*

Different priority processes will misclassify chemicals in different ways. Without a specification of the severity or cost of a misclassification there is no way to compare priority processes and to recommend one over another.

To achieve comparability among alternative priority processes, a concept of misclassification cost is now developed. A misclassification cost (as distinct from a monetary cost) is incurred when a chemical is finally classified as being of a certain type, which is not the true type of the chemical. The final classification of a chemical occurs when, on the basis of the probability distribution associated with the chemical (which reflects the results of the tests to which the chemical has been subjected), the decision is made to terminate testing on it and to label it as a certain type. Write $\hat{s}_i$ to mean that a chemical is classified, by the process, as in category $s_i$. The misclassification of a chemical classified as $\hat{s}_i$ when its true type is $s_j$ is written $C(\hat{s}_i | s_j)$. This cost is zero when $i=j$.

*Budget Cost*

A second requirement for comparability is that the budget cost of a priority process be defined. The expected cost (per chemical) is specified as follows. A priority process $\pi$ is characterized by a set of paths $\pi_i$; $i=1, \ldots M^{\pi}$, where $M^{\pi}$ is the number of paths. Each path corresponds to a sequence of tests and test results.

In the example shown in figure 2, the path $\pi_i$ is the sequence of (test, result) pairs $(k_1,i_1)$, $(k_2,i_2)$, $(k_3,i_3)$, $(k_4,i_4)$.

Considering the test scheme to be one big test, the paths $\{\pi_i; i=1, \ldots M^{\pi}\}$ can be viewed as defining the $M^{\pi}$ test outcomes. As in the case of a simple test, the scheme $\pi$ is characterized by the conditional probabilities: $P(\pi_i | s_j)$.

If the path $\pi_i$ is the sequence $(k_1,i_1)$, $(k_2,i_2)$, …, $(k_n,i_n)$, of tests and results, then

$$P(\pi_i \; s_j) = P(r_{k_1 i_1} | s_j) \, P(r_{k_2 i_1} | s_j) \ldots P(r_{k_n i_n} | s_j).$$

The dollar testing cost $C_{\pi_i}$ associated with a path $\pi_i$ is simply the sum of the costs of the tests along the path, that is

$$C_{\pi_i} = CT_{k_1} + CT_{k_2} + \ldots + CT_{k_n}, \quad \text{where } CT_k \text{ is the cost of test k.}$$

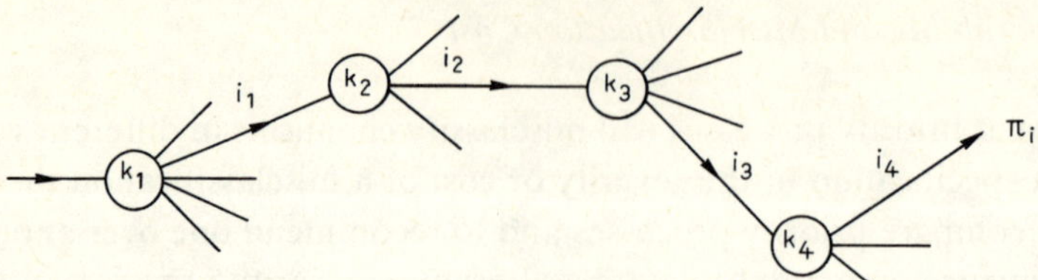


*Fig. 2.* A path with 4 tests.

Rather than a single testing cost, the scheme $\pi$ has a vector of expected costs $\{E_j^\pi; j = 1 \ldots N\}$, since chemicals of different types follow the various paths with different probabilities:

$$E_j^\pi = \sum_{i=1}^{M^\pi} P(\pi_i | s_j) C_{\pi_i} = \sum_{i=1}^{M^\pi} P(\pi_i | s_j)(CT_{k_1} + CT_{k_2} + \ldots CT_{k_n}).$$

The total expected cost ($TCT^\pi$) of process $\pi_i$ per chemical is

$$TCT^\pi = \sum_j^N P(s_j) E_j^\pi.$$

*Minimizing Misclassification Cost*

For an initial probability distribution $P(s_j)$ the probabilities associated with outcome categories of a process $\pi$ are $P(\pi_i | s_j)P(s_j)$; $i = 1, \ldots M^\pi$, $j = 1, \ldots N$.

The best final classification of chemicals in outcome category i is that which minimizes expected misclassification cost. The minimum misclassification cost for path $\pi_i$ is defined as

$$CM_i^\pi = \underset{t=1 \ldots N}{\text{Min}} \sum_j P(\pi_i | s_j) P(s_j) C(\hat{s}_t | s_j),$$

and $t^0$ is the choice of t which minimizes this cost.

The total expected misclassification cost for policy $\pi$ is

$$TCM^\pi = \sum_{i,j} CM_i^\pi P(\pi_i | s_j) P(s_j)$$

per chemical.

The foregoing shows how a priority process $\pi$ acting on an initial collection of chemicals described by a probability distribution $P(s_j)$ incurs a misclassification cost $TCM^\pi$ and a testing cost $TCT^\pi$. Defining the budget limit as B, we now state the priority problem in closed form.

The optimal process is that which minimizes $TCM^\pi$ subject to $TCT^\pi \leq B$.

The above statement of the priority problem has been made operational in terms of a computer program, which chooses the optimal priority process for a given set of parameters. Thus, parameters are: (1) estimates of the distribution of toxicity and exposure over the universe of chemicals $P(s_j)$; (2) the performance characteristics of tests $P(r_{ki} \mid s_j)$; (3) dollar costs of tests $CT_k$, and (4) misclassification costs $C(\hat{s}_i \mid s_j)$.

*Acknowledgement*

We wish to thank the National Science Foundation, the Mellon Foundation, and the National Academy of Sciences for supporting this research.

*References*

1 Astill, B.D.; Lockhart, H.B., Jr.; Moses, J.B.; Nasr, A.N.M.; Raleigh, R.L.; Terhaar, C.J.: Sequential testing for chemical risk assessment; in Conway, Environmental risk analysis of chemicals. 2nd Int. Congr. Toxicol., Brussels 1982 (Van Nostrand-Reinhold Company, New York, in press).

2 Cramer, G.M.; Ford, R.A.; Hall, R.L.: Estimation of toxic hazard. A decision tree approach. Fd Cosmet. Toxicol. *16:* 255–276 (1978).

3 Gori, G.B.: Ranking of environmental contaminants for bioassay priority; in Mohr, Schmahl, Tomatis, Davis, Air pollution and cancer in man, pp. 99–111. IARC Scientific Publications No. 16 (International Agency for Research on Cancer, Lyon 1977).

4 Nisbet, I.C.T.: Ranking chemicals for testing. A priority-setting exercise under the toxic substances control act; in Scoring chemicals for health and ecological effects testing, pp. B-41–B-54. TSCA-ITC Workshop (Enviro Control, Rockville 1979).

5 Weinstein, M.C.: Decision making for toxic substances control. Publ. Policy *27:* 333–383 (1979).

6 Wilhelm, S.: Selecting priorities from large sets of alternatives. The case of toxic substances regulation; Diss. Brown University, Department of Chemistry, Providence, 259 pp. (1981).

T. Page, Environmental Quality Laboratory, California Institute of Technology, Pasadena, CA 91125 (USA)

Safety Evaluation and Regulation of Chemicals. 1st Int. Conf., Boston, Mass., 1982, pp. 112–120 (Karger, Basel 1983)

# Current Status of 'Generally Recognized as Safe' (GRAS) Substances Evaluation

*George W. Irving, Jr.*

Life Sciences Research Office, Federation of American Societies for Experimental Biology, Bethesda, Md., USA

## *History and Definitions*

The Food Additives Amendment of 1958 [3] to the Food, Drug and Cosmetic Act [2] defined food additives[1] and required that each receive Food and Drug Administration's (FDA) prior approval, based on scientific data provided by the petitioner. The amendment also exempted from the requirement of prior approval several hundred substances already in common food use as added ingredients, because they were 'generally recognized as safe' by experts qualified by scientific training and experience to judge safety. A list of these substances, first published by FDA in the early 1960s, came to be called the GRAS (acronym for generally recognized as safe) list and the substances on it, GRAS substances.[2]

Careful scrutiny of the scientific data available on most of them was not undertaken until the FDA was directed to do so by President *Richard M. Nixon* in his 1969 special message to the Congress on consumer protection [6].

The FDA chose to utilize extramural assistance in conducting the evaluation of the safety of the GRAS substances. Under separate contracts, bibliographic searches on each of the substances were made, tests for mutageni-

[1] The amendment defines food additive as any substance the intended use of which results or may reasonably be expected to result, directly or indirectly, in its becoming a component or otherwise affecting the characteristics of any food, including any substance intended for use in producing, preparing, treating, packaging, transporting, or holding food.

[2] The current GRAS list appears as parts 182, 184, and 186 of title 21 of the Code of Federal Regulations [4].

city, carcinogenicity, and teratogenicity were conducted on a selected few of the substances, a survey of the food industry to determine the extent of use of the GRAS substances was undertaken, monographs including all of this information and reprints of the literature on each substance or group of related substances were compiled and evaluation of the total information thus made available was initiated. The evaluation contract, initiated in 1972, was with the Life Sciences Research Office (LSRO) of the Federation of American Societies for Experimental Biology (FASEB).

The evaluation task involved some 400 of the GRAS substances. The remaining substances on the published GRAS list, consisting principally of the natural species and their extractives, were reserved for later and separate consideration. Most of those evaluated were substances that are added directly to food but evaluations were also made of a number of components of food packaging materials which might reach food contained through abrasion or migration. Also included were some 50 substances that were 'prior-sanctioned', that is, were given explicit approval by FDA for specific uses prior to 1958, and a few 'regulated' food additives for which FDA wished evaluations.

### *Evaluation Procedures*

To evaluate the available information on the substances assigned, the LSRO chose to utilize the part-time services of a panel of scientists drawn principally from among nominations made by the six constituent societies of FASEB, with full-time support of the professional and nonprofessional staff of the LSRO. The evaluation panel, named the Select Committee on GRAS Substances (SCOGS)[3] included representatives of the disciplines regarded as necessary to bring expert judgment to bear on the information available on

[3] When work began in 1972, the Select Committee consisted of 9 members but was expanded to 11 in 1974. Of the original 9, Dr. *A.M. Altschul* left the Committee in 1973, and Dr. *S.A. Miller* in 1978. Members at the conclusion of the evaluation effort were: *G.W. Irving*, Jr., PhD, (Chairman), LSRO, FASEB; *J.F. Borzelleca*, PhD, Virginia Commonwealth University; *H.G. Day*, ScD, Indiana University; *S.J. Fomon*, MD, University of Iowa; *B.N. La Du*, Jr., MD, PhD, University of Michigan Medical School; *J.R. McCoy*, VMD, New Jersey College of Medicine and Dentistry; *G.L. Plaa*, PhD, University of Montreal Faculty of Medicine; *M.B. Shimkin*, MD, University of California School of Medicine, San Diego; *R.G.H. Siu*, PhD, Consultant; *M.E. Swendseid*, PhD, University of California School of Public Health, Los Angeles; *J.L. Wood*, PhD, University of Tennessee Center for Health Sciences, Memphis.

each of the substances to be evaluated. While staff members[4] did not share responsibility for decisions of the Select Committee, they participated actively in discussions leading to decisions. Both the Select Committee and staff members were required to be free of associations or activities that would constitute a conflict of interest.

Subcommittees of two or more Select Committee members considered the monographs and other information on each GRAS substance as assigned, and prepared the initial draft report for full Committee review, debate, and revision by staff. The revision process culminated in a tentative report of the Select Committee which was agreed to by all members (except in the two instances where minority reports were found necessary). It was then sent to FDA for announcement in the Federal Register that the Select Committee would entertain requests for a public hearing on the subject of the report. Copies of the tentative report, as well as all of the scientific information on which it was based, were made available for all to examine in the Documents Management Branch of the FDA. Hearings[5] were recorded by a commercial firm from which unedited transcripts could be purchased.

The tentative report, appropriately revised and taking into account information revealed if a hearing was held, became the final report of the Select Committee. This report was reviewed for policy and professional adequacy by the LSRO Advisory Committee, consisting of representatives of each of the constituent societies of FASEB, approved by the Director of LSRO, printed, and submitted to FDA by the Executive Director of FASEB. Subsequently, copies of the report were published by the National Technical Information Service, Springfield, Va., where they may be purchased.

FDA has considered the Select Committee's reports as a major element in making its decision and issuing a proposal in the Federal Register confirming or changing the regulatory status of the substance concerned. To facilitate the decision-making process, it was recognized early that the language of the conclusions in each report should be such that there would be minimal opportunity for misunderstanding the Committee's intent. Accordingly, five conclusion statements were agreed upon as likely to cover all cases. The Committee found that, with a few exceptions, its views could be properly expressed in one or another of the following five statements:

[4] Professional staff of LSRO that participated in all or part of the evaluation of the GRAS substances were: *R.G. Allison,* PhD; *S.A. Anderson,* PhD; *C.J. Carr,* PhD; *H.I. Chinn,* PhD; *S.B. Detwiler,* Jr., MA (deceased); *K.D. Fisher,* PhD; *A.F. Freeman,* BS; *R. Hertz,* MD, PhD; *G.W. Irving,* Jr., PhD; *F.R. Senti,* PhD; *J.M. Talbot,* MD; *M.J. Wade,* PhD.

[5] Hearings were held on 14 of the Committee's tentative reports, covering 77 substances.

(1) There is no evidence in the available information on ..... that demonstrates or suggests reasonable grounds to suspect a hazard to the public when it is used at levels that are now current or that might reasonably be expected in the future.

(2) There is no evidence in the available information on ..... that demonstrates or suggests reasonable grounds to suspect a hazard to the public when it is used at levels that are now current and in the manner now practiced. However, it is not possible to determine, without additional data, whether a significant increase in consumption would constitute a dietary hazard.

(3) While no evidence in the available information on ..... demonstrates a hazard to the public when it is used at levels that are now current and in the manner now practiced, uncertainties exist requiring that additional studies should be conducted.

(4) The evidence on ..... is insufficient to determine that the adverse effects reported are not deleterious to the public health when it is used at levels that are now current and in the manner now practiced.

(5) In view of the deficiency of relevant biological (and/or other) studies, the Select Committee has insufficient data upon which to base an evaluation of ..... when it is used as a food ingredient.

In the long history of the Select Committee, FDA has had no problem in interpreting its conclusions and in translating them into regulatory actions.

It is important to note that in each report, the Select Committee was careful to point out that its conclusions should be reviewed when new information becomes available. Since any report can only reflect the knowledge available at the time of its completion, it is reasonable to expect that reconsideration of the conclusions should be undertaken as new, significant data appear. This was done by the Select Committee, at the request of FDA, several times during the life of the Committee. When the Select Committee will no longer be available for such reconsideration, FDA should consider some other internal or external mechanism for professional evaluation.

*Results*

The health aspects of 468 GRAS or prior-sanctioned substances have been evaluated by the Select Committee and reported to FDA. These include 422 food ingredients used as nutrients, flavoring agents, antioxidants, enzymes, processing aids, or as substances serving other functional purposes. Of these 422, 70.4% were given conclusion No. 1, 16.3% conclusion No. 2, 5.0% conclusion No. 3, 1.2% conclusion No. 4, and 7.1% conclusion No. 5. The remaining 46 substances were evaluated only as components of food packaging materials; 42 of these were given conclusion No. 1 and 4 were given conclusion No. 5. FDA has announced in the Federal Register formal action on about 20% of the substances the Select Committee has evaluated, has proposed action on an additional 25%, and has prepared most of the remainder for Federal Register announcement in the near future. Some observations on the Committee's evaluation experiences follow.

*Relevancy of Biological Data*

Most of the data with which the Select Committee had to work were not generated in experiments designed to shed light on the safety of the substance in question, but were experiments conducted with other objectives in mind. Accordingly, the Committee was expected to exercise reasoned judgment, recognizing early that no fixed standard of adequacy of toxicological data could be generally applied. Each substance was judged in light of its intrinsic biochemical and physiological properties, evidence from available test results involving similar substances, the level of projected use, and the empirical experience of long-term human usage. Where no definitive animal feeding studies had been reported for the substance itself, heavy weight was accorded available toxicological data on major constituents.

*Credibility of Data*

The degree of credibility of reports of scientific studies was found to vary with the nature of the report itself and to some extent with the value judgments of the evaluators as they considered such factors as design and performance of experiments, numbers of observations, use of adequate numbers of experimental animals and controls, relationship between results and conclusions, lack of bias in discussion, availability of confirmatory studies by others, and the reputation of the investigator(s), and/or the reliability of the journal in which the work was published. Initially, the Select Committee relied only on published work, both to avoid the possibility of being misled by unrefereed reports, or by data slanted to a conclusion which might benefit a special interest group whether public or private. However, this practice had to be abandoned in instances where published data were inadequate and where unpublished data were made available at public hearings. With some exceptions, results of investigations contracted for or performed by FDA were usually given the credibility of published reports.

Because of the many possibilities for error in research and in interpretation of results, the Committee learned to view even the most impressive results with caution until verified by others working independently. It also came to regard reliability in a given piece of data as being increased not only by its reproducibility, but in its coherence with other pieces of data within the total available.

*Identity of Food Ingredients – Lack of Quality Control*

While most of the substances added to foods are well characterized and identifiable chemically, many are complex natural products such as gums, starches, proteins, and protein hydrolyzates. These can vary in composition and physical characteristics depending on the source and methods of isolation or preparation.

The need to establish and adhere to food grade standards is obvious for all food ingredients for which such standards have not already been developed and published in the Food Chemicals Codex [5]. With natural plant products especially, the species (and even the general geographic area of its origin) from which a product is derived should be specified as well as any processing or compositional limitations needed to assure safety as demonstrated by biological tests made on the product itself. Once such standards are developed and adopted generally, evaluators need have less doubt than at present that a product for which biological data are available is the substance actually used in foods.

*Consumption Data*

Despite a number of surveys, only incomplete and imprecise consumption data are available for most food ingredients. The Select Committee initially utilized intakes calculated from survey data collected expressly for determining consumption levels of the GRAS substances [8] until experience with them showed them to be rather consistently overestimated. A subsequent NRC survey [1], improved estimates for a limited number of additives. For this reason, the Committee also availed itself of any and all other means for making the estimates necessary, including production, import, and disappearance data and calculations based on known consumption of major foods containing GRAS food ingredients.

Special problems are posed to the evaluator by those substances that (a) are added to foods during processing and subsequently removed, e.g. activated charcoal, tannins: (b) are added for functional purposes but are lost in varying degrees before the food is consumed, e.g. hydrogen peroxide, sulfiting agents; (c) may react with food ingredients to form new substances whose toxicological properties may or may not be known, e.g. sulfiting agents, hydrogen peroxide.

The latter two present the greatest difficulties. Sulfiting agents and hydrogen peroxide are not usually present at the same level in the food as consumed as was added during processing, and can react with and destroy nutrients and create new chemical entities the nature and amounts of which are not always known. This raises two problems for the evaluator. First, it is impracticable to employ data on the quantity of a substance added to food (usually known) for estimating consumer exposure when the substance is present only at an unknown fraction of that amount in the food as consumed. Second, the evaluator must consider, in addition to the biological properties of the added substance itself, the biological properties of identifiable and posssible products of reaction of the added substance with other food ingredients as well as their possible effects on tissue constituents.

All safety evaluations of food ingredients should consider the amount actually present and ingested, not just that added to the food. They should also concern the health aspects of known and potential reaction products between the additive and food constituents, and the possible destruction of nutrients, especially those with which a particular food is identified as a dietary source. Ultimately, residue limits should be established for all substances for which there are concerns about their biological effects.

*Total Intake vs. that Added to Foods*

While the Select Committee's attention was limited to the health aspects of the amounts of GRAS substances added to foods, it was recognized that the safety of a food ingredient for human consumption is a function of the total consumption from all sources, not of just the fraction contributed by commercially processed foods. The Committee found that sometimes the added GRAS substance was the sole or major source of the substance in the diet but more often the substance was already present naturally in foods and the amount added, depending on the level of addition, could make an appreciable or an inappreciable contribution to the total consumed. Obviously, when the evidence indicates that the total consumption is safe, then the fraction attributable to commercial input would also be so regarded. However, where there are questions concerning the health aspects of total consumption, then the practical meaningfulness of limiting the commercially added portion would depend on its proportionate share of the total.

*Concluding Comments*

Some of the experiences of the Select Committee at the time it was about half through the evaluation of the GRAS substances were published in 1977 [7]. Another paper reflecting upon the total experience is expected to be published later this year. Among other things, that paper will include, as an appendix, the conclusions drawn, the call numbers for purchasing copies of the reports, and the references to the Code of Federal Regulation numbers identifying the regulatory category.

The review and assessment of the health aspects of most of the GRAS and prior-sanctioned food ingredients by FDA, with the assistance of many including the Life Sciences Research Office of the Federation of American Societies for Experimental Biology for the evaluation of the available data, has been regarded by many as a wholly satisfactory effort. Accordingly, one may speculate that the procedures used for data collection, selection and operation of the review panels and staff, and conducting public hearings, could be applicable to the evaluation of the health aspects of other food ingredients and possibly other environmental components. Some of the problems that arose in the course of the Select Committee's work are already being considered by FDA as it establishes its positions with respect to proposed changes in food laws [9].

Safety evaluation is a complex process. While the measurement of a parameter in the laboratory can be scientific and objective, the interpretation of the results is open to many subjective and subconscious influences. Members of the Select Committee were acutely aware of the nonscientific components of judgmental deliberations and took special pains in making self, as well as group, examinations from time to time, in order to reduce emotional and idiosyncratic biases in arriving at as sound, collective judgments as possible. Some of the factors to be recognized in this regard and dealt with in forming and operating evaluation panels to minimize the skewing of conclusions, may be considered in the Committee's forthcoming final report.

## *References*

1 Committee on GRAS List Survey – Phase III: 1975 resurvey of the annual poundage of food chemicals generally recognized as safe (GRAS). Prepared under DHEW Contract with Food and Nutrition Board, National Research Council. Rept. PB 288081, NAS (National Technical Information Service, Springfield 1978).

2 Federal Food, Drug and Cosmetic Act as Amended: United States Code, Title 21 (1971).

3 Food Additives Amendment: An act to protect the public health by amending the Federal Food, Drug and Cosmetic Act to prohibit the use in food of additives which have not been tested to establish their safety. 72 STAT 1784 (1958).

4 General Services Administration, Office of the Federal Register: Code of Federal Regulations, Title 21: food and drugs. 21 CFR 182 (US Government Printing Office, Washington 1981).

5 National Research Council: Food Chemicals Codex; 2nd ed. (National Academy of Sciences, Washington 1974).
6 Nixon, R.M.: Special message to the Congress on consumer protection, October 30; in Public Papers of the President's Message 421; 10 pp. (US Government Printing Office, Washington 1969).
7 Select Committee on GRAS Substances: Evaluation of the health aspects of GRAS food ingredients; lessons learned and questions unanswered. Proc. Fed. Am. Soc. exp. Biol. *36:* 2519–2562 (1977).
8 Subcommittee on Review of the GRAS List – Phase II: A comprehensive survey of industry on the use of food chemicals generally recognized as safe (GRAS). Prepared by the Committee on Food Protection, National Research Council. Rept. PB 221920, NAS (National Technical Information Service, Springfield 1972).
9 US Congress: Hatch Bill, No. S1442, Food safety amendment of 1981 (in Congressional Record No. 98, PT. II, June 25, 1981).

G.W. Irving, Jr., PhD, 4836 Langdrum Lane, Chevy Chase, MD 20815 (USA)

Safety Evaluation and Regulation of Chemicals. 1st Int. Conf., Boston, Mass., 1982, pp. 121–125 (Karger, Basel 1983)

# Regulatory Reform: Alternatives to Regulation for Managing Risks to Health, Safety and Environment

*Michael S. Baram*

Boston University School of Law, Boston, Mass., USA

## *Reconsidering Regulation*

Over the last decade, four administrations have sought to reform the regulatory process, by making it more efficient and less burdensome. The fourth administration, President Reagan's, also seeks to go further – to reduce our dependance on federal regulators and to promote private and state government initiatives.

To reach these ends, this administration employs the use of cost-benefit analysis and requires agencies to assess priorities and alternative approaches, under Executive Order 12291 and Office of Management and Budget (OMB) supervision. Executive Order 12291 (February 17, 1981) places great emphasis on the use of cost-benefit analysis in the evaluation and approval of proposed regulations. The President's Office of Management and Budget, and the Vice-President's Regulatory Relief Task Force have been involved with its implementation. These developments succeed various efforts of the Carter Administration [1; also see ref. 2]. Other measures have also been taken, such as drastically reducing agency budgets, cut-backs in agency enforcement, and the appointment of agency officials who are committed to deregulation.

These formal and informal measures, some rational, some ideological, have generated much concern that we, as a nation, will not achieve the level of managing risks to health, safety and environment that is desired by large sectors of the public. These measures have also provoked concern in the corporate sector that state and local initiatives of an infinite variety will be developed to protect health, safety and environment, and that this will impose new burdens and liabilities, and substantially complicate corporate activities across the nation. (For example, whereas the OMB has determined that the proposed chemical labeling regulations of the federal Occupational Safety and Health Administration (OSHA) are not supportable by cost-benefit analysis, and OMB has opposed OSHA; the Chemical Manufacturer's Association has acted in support of the proposed regulation, which it prefers in lieu of multiple legal requirements likely to arise in the states.) The states are also concerned that their hazards will not be dealt with by federal officials, that federal funding for state efforts will be inadequate, and that legal and resource restrictions will obstruct necessary state responses. Finally, researchers in private firms and academia are concerned that calls for their expertise, and federal funds, are now sliding to lower levels.

Thus, the 'new federalism' approach and parallel efforts at regulatory reform of this administration, which bear on health, safety and environmental risks, are beset by these concerns, and the concerns are not capricious but have a rational basis.

It is now apparent that during this time of conflict and change, real opportunities emerge to develop new approaches to managing risks which may be more efficacious and cost-effective, less burdensome and intrusive. These new opportunities will require the assumption of enlarged responsibilities by state and corporate officials. We can call these new policy options, *alternatives to federal regulation* for managing risk to health, safety and environment.

In our research, sponsored by the Ford Foundation and the National Science Foundation, with some results recently published as a book [3], we have been working to identify these alternatives to federal regulation; assess their functions and limitations for risk management; structure their reform so that they can serve as supplements to, or substitutes for, federal regulation; test the alternatives on actual risk problems, and derive the basic principles or criteria to govern their adoption and use. Let me now present some findings.

### *Alternatives to Regulation*

For many years, before and during the inception of regulatory programs for managing health, safety and environmental problems, our society has employed several state and private systems for risk management, with mixed results. These include: private actions at common law (nuisance, negligence, etc.); insurance, trust funds and bonding; industrial self-regulation and professional licensing; negotiation for enforceable agreements between interested parties; tax measures and other economic forces, and federal dissemination of information and procurement restrictions.

Time does not permit discussion of the historical experience with these alternatives, their proven adequacies and limitations, for managing specific types of risks [alternatives fully discussed in ref. 3]. Several of the alternatives are liability systems and impose costs on the corporate sector, after adversary processes in the case of the common law, after consensual processes in the case of the use of insurance. Several are designed to compensate or remedy after the manifestation of risk, and thereby over time can have preventive effects as well. Some permit easy involvement for affected interests, others do not. Some involve considerable cost and expertise, others do not. Some produce efficient and equitable results, others do not. But none are as prescriptive as 'command and control' regulation, and all afford the private sector more flexibility in fashioning more finely-tuned and cost-effective approaches to risk management.

For example, reliance on common law leads to different considerations in each of the 50 states, requires that plaintiffs bear up-front costs for lawyers and experts, is hindered by short statutes of limitations and other defenses available to well-endowed corporate defendants. But many common law reforms are taking place, and now class actions and other measures expedite use of common law litigation. Historically, the common law has been dynamic and responded to meet societal needs for risk management. Thus, if we choose increasingly to rely on the common law to manage risk, basic reforms will have to be undertaken. So the opportunity exists to install a reformed and effective and dynamic common-law approach to risk management, and this may prove to be more useful and acceptable over time than ever-growing federal regulation. Other opportunities also exist to develop other improved nonregulatory approaches, but time does not permit discussion here.

## *The Case of Hazardous Waste Management*

To try to bring some of the foregoing concepts to practice, consider the case of hazardous waste management. Despite the enactment of RCRA in 1976 [4], and the promulgation of thousands of pages of EPA regulations for 'cradle to grave' management of such wastes (40 CFR, Parts 261–267, for

Environmental Protection Agency Regulations), which, incidentally, are still incomplete and also being continuously reconsidered, we still face enormous hazardous waste problems.

Let us briefly consider some of the alternatives. Why not permit each state to develop its own approach involving a balanced and integrated set of strategies which could include, for example, an improved common law incorporating strict liability principles and new law for the protection of ground water; new tax measures to force reduction of waste streams, and to provide incentives for incineration or treatment systems, rather than landfill; industrial insurance requirements, with rates keyed to risk and loss experience; negotiated agreement frameworks for all affected interests to deal with the potential social costs associated with the siting, maintenance and closure of generator and disposal facilities.

These options could serve as supplements to EPA regulation and obviate the need for extensive further regulation by the agency [5]. They could also serve as substitutes for much EPA regulation, unless their application were subsequently found to be inadequate. This approach will require careful evaluation before adoption, and certain criteria could be employed by each state and EPA in forging this new approach. The criteria could include considerations of efficiency (more cost-effective than regulation); equity (in terms of fair distribution of costs and benefits, and provision of due process to interested parties); evidentiary feasibility (that evidentiary requirements are feasible); institutional simplicity and adequacy of authority; public acceptability, and efficacy, overall.

Although factual findings of a conclusive nature for each of these criteria to be applied to each alternative are impossible to achieve, a mixed set of facts and opinions can be developed in order to determine whether each criterion can be met by the proposed alternative.

The residual federal role in this new regime of alternatives would involve providing research and findings to potential users; providing focussed expertise to litigants or others in actual cases; monitoring state developments and reinstituting federal regulations as required, and brokering harmonized state programs, to prevent gross discrepancies between states which leads to out-of-state dumping, and to prevent undue restrictions on interstate commerce.

Thus, we believe a gradual reduction of hazardous waste regulations can be properly conducted, and the use of alternatives can prove to be efficacious, efficient and acceptable. Similar options exist in other risk sectors which involve consumer protection, worker health, and groundwater quality, for example.

## *Conclusions*

It is unlikely that any federal agencies will emerge from the current period of regulatory reform with a renewed mandate for comprehensive regulation. Thus, it is imperative that we now focus on the development of sound and acceptable alternatives to regulation for managing risks.

## *References*

1 US Regulatory Council: Regulating with common sense (1980).
2 Baram, M.: Regulation of health, safety and environment and the use of cost-benefit analysis. Report to the Administrative Conference of the United States, Washington 1979.
3 Baram, M., et al.: Alternatives to regulation. Lexington Books (Heath, Lexington 1981).
4 Resource Conservation and Recovery Act of 1976, 42 USC §6901.
5 Baram, M.: Expanding the policy options for managing hazardous wastes. Report to the Office of Technology Assessment, US Congress (December 1981).

Prof. Michael S. Baram, Boston University Law School, Room 1620,
765 Commonwealth Avenue, Boston, MA 02215 (USA)

# In vitro Testing: Mutagenesis, Carcinogenesis

Safety Evaluation and Regulation of Chemicals. 1st Int. Conf., Boston, Mass., 1982, pp. 126–133 (Karger, Basel 1983)

## Bacterial Mutagenicity Testing: Aims, Achievements, and Limitations

*Michael J. Prival*

Genetic Toxicology Branch, Food and Drug Administration, Washington, D.C., USA

### *The Ames Salmonella Test*

The most widely used bacterial test for mutagenicity, and the most widely used short-term test for carcinogenicity, is the 'Ames' Salmonella test. The popularity of this test for screening chemicals for potential carcinogenicity has several reasons. First, results from several independent large-scale studies have shown that there is a reasonably good correlation between mutagenicity in Salmonella and carcinogenicity in mammals [24, 25, 36, 40, 43]. In addition, the Ames test is relatively rapid, simple, and inexpensive to perform compared with other tests for genetic or carcinogenic effects.

Many investigators believe that the Ames test is useful for detecting carcinogens because chemicals initiate the carcinogenic process by inducing mutations in somatic cells. Although it appears likely that the initiation of carcinogenesis may involve somatic mutation, this is still a matter of some dispute. The utility of the Ames test and other mutagenesis tests in detecting carcinogens does not rest upon any somatic mutation theory of carcinogenesis, but rather upon the fact that carcinogens tend to be mutagens and vice versa. Thus, whatever the mechanistic basis for the correlation between carcinogenesis and mutagenesis, it is the empirical observation of this correlation, rather than mechanistic concepts, that underlies the utility of mutagenesis tests in screening chemicals for potential carcinogenicity.

The correlation between carcinogenicity and mutagenicity will vary greatly between chemicals of different classes, which makes it impossible to state that there is a particular 'percent correlation' between the results in the Ames Salmonella assay and whole animal, chronic carcinogenesis tests. The 'percent correlation' is also difficult to establish because the whole animal bioassays are, themselves, often difficult to interpret in terms of carcinogenicity.

Because of these uncertainties, it is difficult to come up with specific percentages to indicate the specificity or sensitivity of bacterial mutagenicity tests. The utility of such tests is best understood by considering specific cases that have led us to ask important questions concerning the safety of chemicals to which humans have been widely exposed.

*Case Studies*

There is at least one case in which a heavily used food additive was identified first as a bacterial mutagen and later as a carcinogen. In 1965, the nitrofuran compound furylfuramide or AF-2 was approved for use as a preservative in foods by the Japanese government. Long-term testing on rats and mice indicated that this chemical was not a carcinogen when fed at levels up to 0.2% in the diet [26]. Since so many nitrofuran compounds are mutagenic and carcinogenic [14], it is really not suprising that AF-2 was found to be a bacterial mutagen [17, 21, 52]. By 1974, evidence for the mutagenicity of AF-2 had also been obtained in Neurospora and yeast [32], silkworms [46], and human lymphocytes in culture [47]. Because of the concern generated by the broad mutagenic activity of this food additive, AF-2 was retested in a chronic animal bioassay in mice and found to be carcinogenic [16, 48, cit. in 20, 31, 38, 44]. As a result of these tests, AF-2 was banned as a food preservative in Japan, but only after about a decade of use in a variety of widely consumed Japanese foods.

Bacterial mutagenicity testing has also been useful in the safety evaluation of flame retardants. Tris(2,3-dibromopropyl)phosphate, or 'tris-BP', was used in most of the children's sleepwear in the United States. This flame-retardant chemical was added to the fabric at levels often exceeding 5% of the weight of the material itself. After several years of widespread use, tris-BP was tested in the Ames Salmonella assay and found to be mutagenic [4, 34]. As a result of the findings in bacteria, this compound was tested in a variety of other biological assays and was found to cause heritable genetic damage in *Drosophila melanogaster* [3, 49], as well as repairable breaks in DNA [13] and unscheduled DNA synthesis in cultured human cells [41]. The results of these genetic tests were then confirmed by experiments demonstrating the carcinogenicity of tris-BP [29, 37, 50]. In addition, experiments conducted at the U.S. Consumer Product Safety Commission laboratories showed that dermal application of tris-BP to male rabbits resulted in testicular atrophy and chronic interstitial nephritis [33]. Further experiments by the Consumer Product Safety Commission showing that labeled tris-BP can be absorbed through the skin from fabric led to the banning of this flame retardant [9].

After tris-BP was banned, industry introduced a replacement for this compound. The replacement is tris(1,3-dichloro-2-propyl)phosphate, or 'tris-CP' (also known as Fyrol FR-2), which is closely related in chemical structure to tris-BP. When tris-CP was tested in the standard Ames assay, it was nonmutagenic [34]. This assay was performed with an activation system derived from rats that had been treated with Aroclor 1254 ( a mixture of polychlorinated biphenyls) to induce liver microsomal enzymes, as is standard practice [1]. However, when the liver S-9 was derived from phenobarbital-induced rats, mutagenic activity could be detected [7, 11]. In general, tris-CP appears not to be as active genetically as is tris-BP [7], which may indicate that it is safer when used as a flame retardant, but its structure and potential for human exposure indicate that it should be tested for carcinogenicity in whole animals. Such testing has apparently not yet been performed [15].

Hair dye ingredients are another group of chemicals that has received a great deal of attention because of mutagenicity in bacteria. Most hair dye products are complex mixtures containing, among other things, aromatic amines. In 1975, it was reported that most commercially available hair dyes are mutagenic [2, 39]. *Ames* et al. [2] found that one of the mutagenic ingredients in hair dyes was 4-methoxy-*m*-phenylenediamine (4-MMPD; 2,4-diaminoanisole). By 1978, results of a feeding study conducted for the National Cancer Institute showed that feeding of 4-MMPD resulted in the induction of thyroid tumors in rats and mice [30, 51]. Because of these findings and evidence that 4-MMPD can be absorbed through human skin [22], the Food and Drug Administration issued a regulation requiring that hair dyes containing 4-MMPD be labeled to indicate that they contain a carcinogen [10]. The use of 4-MMPD in hair dyes was voluntarily halted, but at least one company then introduced another closely related compound, 4-ethoxy-*m*-phenylenediamine (4-EMPD). In view of the close structural similarity between 4-MMPD and 4-EMPD, it is not surprising that their mutagenic activities were found to be similar [35]. Apparently, 4-EMPD has now been voluntarily withdrawn from hair dyes because of its mutagenic activity.

The fact that the Ames test was first used to detect the mutagenicity of hair dyes and then to determine which of the ingredients in these commercial dye mixtures were genetically active illustrates the utility of this test in the analysis of complex mixtures. Another good example is shown in work done on the mutagenicity of lipsticks. When *Muzzall and Cook* [27] found that some lipsticks were mutagenic, they carefully analyzed the ingredient labels on the products they had studied. They discovered that all of the lipsticks bearing labels stating that they contained the dye D&C Orange No. 17 were mutagenic; those that did not contain the dye were not mutagenic. Some lipsticks were labeled that they 'may contain' D&C Orange No. 17; some of these were mutagenic and others were not. *Muzzall and Cook* therefore tested this dye and found that it was in fact mutagenic. They also found that one of the azo reduction products of this dye, 2,4-dinitroaniline, which would be formed in the intestine upon ingestion of the dye, was also mutagenic. *Green and Pastewka* [12] independently made the same observations and drew the same conclusion concerning the origin of the mutagenic activity of lipsticks.

Food is one of the most important types of mixtures that has been found to be mutagenic in the Ames test. Some of the most potent mutagenic substances known can be produced by the pyrolysis of individual amino acids, such as tryptophan, phenylalanine, lysine, and glutamic acid. Several years ago, widespread publicity was given to a report by *Commoner* et al. [8] at Washington University in St. Louis that the cooking of hamburgers resulted in the formation of mutagens. The mutagenic substances responsible for this activity are now known to include 2-amino-3-methylimidazo[4,5-*f*]quinoline (IQ) and 2-amino-3,8-dimethyl-imidazo[4,5-*f*]quinoxaline (MeIQx) [18, 19]. Both IQ and the mutagen 2-amino-3,4-dimethylimidazo[4,5-*f*]quinoline (MeIQ) have been isolated from broiled fish [18]. Two mutagens, produced when tryptophan is pyrolyzed, induced hepatocellular carcinomas when fed to mice at 200 ppm in the diet [23]. In addition, mutagens formed during the pyrolysis of glutamate or soybean globulin are also reportedly carcinogenic, although the data have not yet been published [42]. Further testing of these and other mutagenic pyrolysis products of foods is essential if we are to develop some understanding of the possible significance of cooking in adding to our exposure to chemical carcinogens. In the interim, all that can be said is that such mutagens appear to be present at very low levels, if at all, in normally cooked, uncharred foods. Since the cooking of foods can destroy pathogenic micro-organisms and parasites, it is unwise at this time to recommend that the amount of cooking of meat or other foods be reduced.

Another widespread class of mutagens in foods is not produced by cooking but is present as a natural component of many plants. This class of compounds is the flavonoids, as exemplified by the mutagenic flavonol quercetin. Flavonols often exist in plants in the form of glycosides. For example, rutin is a common glycoside of quercetin. When eaten, glycosides such as rutin will be hydrolyzed in the intestinal tract to release their mutagenic aglycones, such as quercetin. Although quercetin and other mutagenic flavonoids are not very potent as mutagens, they are widely distributed in edible plants and are present in considerable quantities. For example, *Brown* [5] has used 50 mg/day as a very rough estimate of the amount of quercetin equivalents consumed daily by humans in the form of promutagenic flavonol glycosides.

Normally, under the conditions that exist in the Ames Salmonella assay, significant hydrolysis of glycosides will not occur, although such hydrolysis does take place when glycosides are consumed by humans or other mammals. To overcome this deficiency in the in vitro metabolic activation system, glycosidases extracted from various sources, such as the mold *Aspergillus niger* [28], the snail *Helix pomatia* [6], the cecal contents of rats [6], or human feces [45], can be added. Similarly, the addition of enzymes capable of hydrolyzing glucuronides and sulfate esters has been used to free conjugated mutagens in the testing of urine from treated animals. The addition of such enzymes to release mutagens from esters and glycosides is one example of how the Ames test must sometimes be modified to permit detection of the mutagenic activity of certain classes of chemicals.

### *Conclusion*

Testing chemicals for mutagenic activity in bacteria appears to be a useful component of a safety evaluation program. In a number of the cases that have been discussed, such as AF-2, tris-CP, and 4-EMPD, positive results in bacterial mutagenicity tests have led to concern regarding the safety of the chemical. In each of these cases, however, the clear structural similarity between the chemical and known carcinogens and mutagens should have been an adequate indication that further safety evaluation in whole animals was required before widespread human exposure could be justified. Such consideration of structure-effect relationships should have made the bacterial mutagenesis test results irrelevant. That is, even if they were shown to be nonmutagenic to bacteria, their chemical structures still would indicate a need for in vivo testing.

In summary, mutagenicity testing in bacteria, especially the Ames Salmonella assay, has proven to be useful in drawing attention to environmental mutagens and carcinogens. Many mutagens and carcinogens in certain chemical classes require specific modifications of the standard Ames test procedure in order for their activities to be expressed. Therefore, negative results in a routine Ames test are alone insufficient to indicate safety. Negative results in a variety of test systems increase our confidence that a chemical is actually nonmutagenic. However, if all of the test systems rely upon similar in vitro metabolic activation systems, negative results in all tests may simply reflect the limitations of in vitro activation rather than the safety of the chemical.

## *References*


1 Ames, B.N.; McCann, J.; Yamasaki, E.: Methods for detecting carcinogens and mutagens with the Salmonella/mammalian microsome mutagenicity test. Mutation Res. *31:* 347–364 (1975).

2 Ames, B.N.; Kammen, O.; Yamasaki, E.: Hair dyes are mutagenic. Identification of a variety of mutagenic ingredients. Proc. natn. Acad. Sci. USA *72:* 2423–2427 (1975).

3 Berkowitz, S.: The induction of II–III translocations by tris-(2,3-dibromopropyl) phosphate in Drosophila. Mutation Res. *57:* 385–387 (1978).

4 Blum, A.; Ames, B.N.: Flame-retardant additives as possible cancer hazards. Science *195:* 17–23 (1977).

5 Brown, J.P.: A review of the genetic effects of naturally occurring flavonoids, anthraquinones and related compounds. Mutation Res. *75:* 243–277 (1980).

6 Brown, J.P.; Dietrich, P.S.: Mutagenicity of plant flavonols in the Salmonella/mammalian microsome test. Activation of flavonol glycosides by mixed glycosidases from rat cecal bacteria and other sources. Mutation Res. *66:* 233–240 (1979).

7 Brusick, D.; Matheson, D.; Jagannath, D.R.; Goode, S.; Lebowitz, H.; Reed, M.; Roy, G.; Benson, S.: A comparison of the genotoxic properties of tris(2,3-dibromopropyl) phosphate and tris(1,3-dichloro-2-propyl) phosphate in a battery of short-term bioassays. J. envir. Pathol. Toxicol. *3:* 207–226 (1980).

8 Commoner, B.; Vithayathil, A.J.; Dolara, P.; Nair, S.; Madyastha, P.; Cuca, G.C.: Formation of mutagens in beef and beef extract during cooking. Science *201:* 913–916 (1978).

9 Federal Register: *42:* 18850–18854 (April 8, 1977).

10 Federal Register: *44:* 59509–59522 (October 16, 1979).

11 Gold, M.D.; Blum, A.; Ames, B.N.: Another flame retardant, tris-(1,3-dichloro-2-propyl)-phosphate, and its expected metabolites are mutagens. Science *200:* 785–786 (1978).

12 Green, M.R.; Pastewka, J.V.: Mutagenicity of some lipsticks and their dyes. J. natn. Cancer Inst. *64:* 665–669 (1980).

13 Gutter, B.; Rosenkranz, H.S.: The flame retardant tris(2,3-dibromopropyl) phosphate. Alteration of human cellular DNA. Mutation Res. *56:* 89–90 (1977).

14 IARC: IARC monographs on the evaluation of carcinogenic risk of chemicals to man, vol. 7, pp. 143–196 (International Agency for Research on Cancer, Lyon 1974).

15 IARC: Information bulletin on the survey of chemicals being tested for carcinogenicity, No. 9 (International Agency for Research on Cancer, Lyon 1981).

16 Ikeda, Y.; Horiuchi, S.; Furuya, T.; Uchida, O.; Suzuki, K.; Azegami, J.: Induction of gastric tumors in mice by feeding of furylfuramide (In Japanese). Food Sanitation Study Council, Ministry of Health and Welfare, Japan 1974.

17 Kada, T.: *Escherichia coli* mutagenicity of furylfuramide. Jap. J. Genet. *48:* 301–305 (1973).

18 Kasai, H.; Yamaizumi, Z.; Wakabayashi, K.; Nagao, M.; Sugimura, T.; Yokoyama, S.;

Miyazawa, T.; Spingarn, N.E.; Weisburger, J.H.; Nishimura, S.: Potent novel mutagens produced by broiling fish under normal conditions. Proc. Jap. Acad. *56(B):* 278–283 (1980).

19 Kasai, H.; Yamaizumi, Z.; Shiomi, T.; Yokoyama, S.; Miyazawa, T.; Wakabayashi, K.; Nagao, M.; Sugimura, T.; Nishimura, S.: Structure of a potent mutagen isolated from fried beef. Chem. Lett. 485–488 *(1981).*

20 Kinebuchi, M.; Kawachi, T.; Matsukura, N.; Sugimura, T.: Further studies on the carcinogenicity of a food additive, AF-2, in hamsters. Fd Cosmet. Toxicol. *17:* 339–341 (1979).

21 Kondo, S.; Ichikawa-Ryo, H.: Testing and classification of mutagenicity of furylfuramide in *Escherichia coli.* Jap. J. Genet. *48:* 295–300 (1973).

22 Marzulli, F.N.; Green, S.; Maibach, H.I.: Hair dye toxicity. A review. J. envir. Pathol. Toxicol. *1:* 509–530 (1978).

23 Matsukura, N.; Kawachi, T.; Morino, K.; Ohgaki, H.; Sugimura, T.; Takayama, S.: Carcinogenicity in mice of mutagenic compounds from a tryptophan pyrolyzate. Science *213:* 346–347 (1981).

24 McCann, J.; Ames, B.N.: Detection of carcinogens as mutagens in the *Salmonella*/microsome test. Assay of 300 chemicals. (Discussion). Proc. natn. Acad. Sci. USA *73:* 950–954 (1976).

25 McCann, J.; Choi, E.; Yamasaki, E.; Ames, B.N.: Detection of carcinogens as mutagens in the *Salmonella*/microsome test. Assay of 300 chemicals. Proc. natn. Acad. Sci. USA *72:* 5135–5139 (1975).

26 Miyaji, T.: Acute and chronic toxicity of furylfuramide in rats and mice. Tokohu J. exp. Med. *103:* 331–369 (1971).

27 Muzzall, J.M.; Cook, W.L.: Mutagenicity test of dyes used in cosmetics with the Salmonella/mammalian microsome test. Mutation Res. *67:* 1–8 (1979).

28 Nagao, M.; Morita, N.; Yahagi, T.; Shimizu, M.; Kuroyanagi, M.; Fukuoka, M.; Yoshihira, K.; Natori, S.; Fujino, T.; Sugimura, T.: Mutagenicities of 61 flavonoids and 11 related compounds. Envir. Mutagen. *3:* 401–419 (1981).

29 National Cancer Institute: Bioassay of tris (2,3-dibromopropyl) phosphate for possible carcinogenicity. Technical Report Series No. 76, DHEW Publication No. (NIH) 78–1326 (National Cancer Institute, Bethesda 1978).

30 National Cancer Institute: Bioassay of 2,4-diaminoanisole sulfate for possible carcinogenicity. Technical Report Series No. 84, DHEW Publication No. (NIH) 78–1334 (National Cancer Institute, Bethesda 1978).

31 Nomura, T.: Carcinogenicity of the food additive furylfuramide in foetal and young mice. Nature, Lond. *258:* 610–611 (1975).

32 Ong, T.; Shahin, M.M.: Mutagenic and recombinogenic activities of the food additive furylfuramide in eukaryotes. Science *184:* 1086–1087 (1974).

33 Osterberg, R.E.; Bierbower, G.W.; Hehir, R.M.: Renal and testicular damage following dermal application of the flame retardant tris(2,3-dibromopropyl) phosphate. J. Toxicol. envir. Hlth *3:* 979–987 (1977).

34 Prival, M.J.; McCoy, E.C.; Gutter, B.; Rosenkranz, H.S.: Tris(2,3-dibromopropyl) phosphate: Mutagenicity of a widely used flame retardant. Science *195:* 76–78 (1977).

35 Prival, M.J.; Mitchell, V.D.; Gomez, Y.P.: Mutagenicity of a new hair dye ingredient. 4-ethoxy-*m*-phenylenediamine. Science *207:* 907–908 (1980).

36 Purchase, I.F.H.; Longstaff, E.; Ashby, J.; Styles, J.A.; Anderson, D.; Lefevre, P.A.;

Westwood, F.R.: An evaluation of 6 short-term tests for detecting organic chemical carcinogens. Br. J. Cancer *37:* 873–959 (1978).

37 Reznik, G.; Reznik-Schüller, H.M.; Rice, J.M.; Hague, B.F., Jr.: Pathogenesis of toxic and neoplastic renal lesions induced by the flame retardant tris(2,3-dibromopropyl)-phosphate in F344 rats, and development of colonic adenomas after prolonged oral administration. Lab. Invest. *44:* 74–83 (1981).

38 Sano, T.; Kawachi, T.; Matsukura, N.; Sasajima, K.; Sugimura, T.: Carcinogenicity of a food additive, AF-2, in hamsters and mice. Z. Krebsforsch. *89:* 61–68 (1977).

39 Searle, C.E.; Harnden, D.G.; Venitt, S.; Gyde, O.H.B.: Carcinogenicity and mutagenicity tests of some hair colourants and constituents. Nature, Lond. *255:* 506–507 (1975).

40 Simmon, V.F.: In vitro mutagenicity assays of chemical carcinogens and related compounds with *Salmonella typhimurium.* J. natn. Cancer Inst. *62:* 893–899 (1979).

41 Stich, H.: Personal commun. Cited in *Blum and Ames* [6].

42 Sugimura, T.: A view of a cancer researcher on environmenal mutagens. 3rd Int. Conf. Envir. Mutagens, Tokyo 1981.

43 Sugimura, T.; Sato, S.; Nagao, M.; Yahagi, T.; Matsushima, T.; Seino, Y.; Takeuchi, M.; Kawachi, T.: Overlapping of carcinogens and mutagens; in Magee et al., Fundamentals in cancer prevention, pp. 191–215 (University Park Press, Baltimore 1976).

44 Takayama, S.; Kuwabara, N.: The production of skeletal muscle and mammary tumors in rats by feeding 2-(2-furyl)-3-(5-nitro-2-furyl)-acrylamide. Toxicol. Lett. *1:* 11–16 (1977).

45 Tamura, G.; Gold, C.; Ferro-Luzzi, A.; Ames, B.N.: Fecalase. A model for activation of dietary glycosides to mutagens by intestinal flora. Proc. natn. Acad. Sci. USA *77:* 4961–4965 (1980).

46 Tazima, Y.: Some problems in environmental mutagens. From the experiences on AF-2 (In Japanese). Kagaku *43:* 745–749 (1973).

47 Tonomura, A.; Sasaki, M.S.: Chromosome aberrations and DNA repair synthesis in cultured human cells exposed to nitrofurans. Jap. J. Genet. *48:* 291–294 (1973).

48 Uchida, Y.; Furiya, T.; Kawamata, K.; Horiuchi, S.; Ikeda, Y.: Study about toxicity of furylfuramide (AF-2). Second report. Study of carcinogenicity (In Japanese). Folia pharmacol. jap. *71:* 80–81 (1975).

49 Valencia, R.: Drosophila mutagenicity tests for saccharin, tris, $PtCl_4$ and other compounds. 9th Ann. Meet. Envir. Mutagen Soc., San Francisco 1977.

50 Van Duuren, B.L.; Loewengart, G.; Seidman, I.; Smith, A.C.; Melchionne, S.: Mouse skin carcinogenicity tests of the flame retardants tris(2,3-dibromopropyl)phosphate, tetrakis (hydroxymethyl)phosphonium chloride, and polyvinyl bromide. Cancer Res. *38:* 3236–3240 (1978).

51 Ward, J.M.; Stinson, S.F.; Hardisty, J.F.; Cockrell, B.Y.; Hayden, D.W.: Neoplasms and pigmentation of thyroid glands in F344 rats exposed to 2,4-diaminoanisole sulfate, a hair dye component. J. natn. Cancer Inst. *62:* 1067–1070 (1979).

52 Yahagi, T.; Nagao, M.; Hara, K.; Matsushima, T.; Sugimura, T.; Bryan, G.T.: Relationships between the carcinogenic and mutagenic or DNA-modifying effects of nitrofuran derivatives, including 2-(2-furyl)-3-(5-nitro-2-furyl)acrylamide, a food additive. Cancer Res. *34:* 2266–2273 (1974).

M.J. Prival, PhD, Food and Drug Administration, 200 C Street SW,
Washington, DC 20204 (USA)

Safety Evaluation and Regulation of Chemicals. 1st Int. Conf., Boston, Mass., 1982, pp. 134–141 (Karger, Basel 1983)

# Evaluation of Mutagenic and Carcinogenic Potential Using Multiple Genetic Endpoints in Cultured Mammalian Cells

*D.L. Putman, L.M. Schechtman, R.D. Curren, R.E. Kouri*

Division of Toxicology/Oncology, Microbiological Associates, Bethesda, Md., USA

Each year numerous chemical agents are synthesized, many of which may affect the human population upon occupational or environmental exposure. One of the possible adverse effects of exposure to hazardous substances is an alteration in the DNA molecule, with subsequent risk of heritable mutation or carcinogenesis. In order to detect the mutagenic or carcinogenic potential of a variety of chemical agents, a battery of testing procedures employing different target cells, genetic endpoints, and molecular mechanisms has been employed in screening programs. Of the test systems available, cultured mammalian cells have been utilized for a number of reasons. In addition to being economical with regard to cost, time, and technical expertise required, mammalian cell systems fill the gap between the frequently used bacterial assays and the whole animal, short-term bioassays.

Like the bacterial assays, the in vitro mammalian systems use relatively high numbers of target cells which give an advantage for detection of a rare event, and permit a high degree of control regarding exposure conditions. Furthermore, the relevance of a mammalian system including intact enzyme systems and chromosome structures is maintained. Cultured mammalian cells also possess a versatility which allows for a number of genetic events induced by the same or different DNA lesions to be evaluated using the same cell type. This capability greatly increases the sensitivity of such a system by increasing its ability to identify potential genetic hazards.

Several mammalian cell target systems have been developed for the routine assessment of mutagenic or carcinogenic potential of test chemicals. Of these, only a few can be used for evaluation of multiple genetic endpoints, such as point mutations, chromosomal aberrations, primary DNA damage, and cell transformation. Some of the mammalian cell systems which can be

*Table I.* Multiple genetic endpoints in cultured mammalian cells

| Cell type | Gene mutations | Chromosome aberrations | Primary DNA damage | Morphological transformation |
|---|---|---|---|---|
| Chinese hamster lung (V79) | + | + | + | |
| Chinese hamster ovary (CHO) | + | + | + | |
| Mouse lymphoma L5178Y | + | + | + | |
| Syrian hamster embryo | + | + | + | + |
| BALB/3T3 mouse embryo | + | | | + |
| C3H 10T 1/2 mouse embryo | + | | | + |

utilized in studies evaluating one or more of these genetic events are listed in table I. Other systems are currently being developed and validated for multiple endpoint studies; however, their inclusion in screening programs seems premature at this time.

### *Multiple Mutation Endpoints*

A number of genetic loci have been characterized for the study of gene mutations in mammalian cells; however, the three loci most commonly employed in short-term tests include those coding for hypoxanthine guanine phosphoribosyl transferase (HGPRT), thymidine kinase (TK), and the membrane receptor determining ouabain resistance ($OUA^R$). Several authors have reported differences in the sensitivity of various loci to different chemicals. For example, *Carver* et al. [2] have found considerable differences in the response of four genetic loci (HGPRT, TK, $OUA^R$, and adenine phosphoribosyltransferase) in CHO cells to a variety of chemicals.

In addition to monitoring gene mutations at a number of loci, the established near-diploid rodent cell lines can also be used to evaluate the induction of chromosome aberrations, both numerical and structural, and primary DNA damage, such as sister chromatid exchanges (SCEs). Multiple endpoint studies allow for a comparison of minimal effective concentrations of compounds with similar structure and provide a basis for prediction of presumed genetic risk. *Connell* [3] compared the relative potencies of three ultimate metabolites of benzo(a)pyrene and of 7-bromomethybenz(a)anthracene in the induction of gene mutations at the HGPRT locus, SCEs, and chromosome aberrations in V79 cells, and concluded from these studies that gene mutations, chromosome mutations, and SCEs are independent events. *Brusick* et al. [1] have compared the genotoxic properties of two flame retardants, tris(2,3-dibromopropyl)phosphate (TBPP) and tris(1,3-dichloro-2-propyl)phosphate (TCPP) using point mutation at the TK locus, chromosome aberrations, and SCEs in the mouse lymphoma cell line, L5178Y. TBPP was found to be positive in all three assays while TCPP was only weakly positive for the induction of chromosome aberrations. Comparison of minimal effective concentrations also demonstrates a difference in sensitivity of these assays for TCPP; SCEs were induced at 0.005 μl/ml, chromosome aberrations at 0.01 μl/ml and mutations at the TK locus at 0.025 μl/ml. These observed differences substantiate the need for multiple independent endpoints in screening programs in order to limit false-negative results.

### *Simultaneous Mutation and Transformation*

Transformation assays allow for the in vitro evaluation of carcinogenic potential by a system believed to be very similar to the in vivo process. The most frequently used endpoint when using fibroblast cells is a morphological transformation characterized by misoriented, criss-crossed piling up of cells resulting in an aberrant colony or focus. Subsequent passage of these cells allows for the expression of other characteristic phenotypes which include enhanced fibrinolytic activity, growth in soft agar and tumor formation in syngeneic or immunosuppressed hosts. Normal diploid Syrian hamster embryo (SHE) cells and the aneuploid established rodent cell lines, such as BALB/3T3 and C3H 10T 1/2, comprise the major transformable cell lines which can also be utilized for the simultaneous evaluation of mutagenic potential.

The transformation assay using BALB/3T3 cloned cell lines was designed to allow the visualization of transformed foci of high cell density and aberrant cell morphology on a confluent monolayer of nontransformed, contact-inhibited 3T3 cells [5,6]. The BALB/3T3 clone A31-1 cell line [6] derived from the aneuploid, nontumorigenic BALB/3T3 clone A31 cell is characterized by a cloning efficiency of usually greater than 30%, postconfluence inhibition of proliferation, and the ability to form distinct foci of aberrant cell morphology following exposure to chemical carcinogens. Clone A31-1 can also be scored for transformed foci using the criteria established by *Reznikoff* et al. [7, 8]. Our laboratory has used the BALB/3T3 clone A31-1 cell line for simultaneous evaluation of morphological transformation and mutation at the $OUA^R$ locus. We have demonstrated that 3T3 cells can respond in a dose-dependent fashion for both mutagenesis and transformation to N-methyl-N′-nitro-N-nitrosoguanidine (MNNG) in the absence of exogenous metabolic activation and benzo(a)pyrene (BaP) in the presence of exogenous metabolic activation [4, 9]. However, the value of simultaneous transformation and mutation assays is better demonstrated by studies in which dose responses are not obtained or by studies in which only one of the two endpoints is clearly positive. In a study of particulate extracts from a number of environmental emission sources [4], we found no clear dose-dependent responses although several of the samples showed significant transforming and mutagenic activity. As shown in table II, a selected gasoline exhaust extract induced a substantial increase in type III foci at a single dose, 10 μg/ml. Substantial mutagenic activity was found at 10 and 2.5 μg/ml. Similar responses were also found for gasoline emission extracts and coke oven concentrates when tested in the presence and absence of an exogenous source of metabolic activation. In circumstances such as these, a positive response in both assays, in the absence of a clear dose response in either assay, greatly increases confidence in the data.

Additional studies in our laboratory using 6-aminochrysene (6-AC) reinforce the importance of simultaneous mutation and transformation procedures. 6-AC was found to be mutagenic but not transforming when tested in BALB/3T3 cells (table III). Although ouabain-resistant mutants were induced by a 24-hour treatment of BALB/3T3 cells with either 2.5, 5.0 or 10.0 μg 6-AC/ml, identical treatments failed to induce morphologically transformed foci. It was reasoned that 6-AC may induce a different or delayed transformation response resulting in phenotypic lag in the morphological expression of the transformation event. However, allowing a longer expression period by simply continuing the time in culture from 4 to 8 weeks did not result in observable transformed foci. A level II or amplification assay was then performed using 6-AC and BALB/3T3 cells. The primary difference between the standard assay and the level II assay is the introduction of a single subcultivation of target cells after a 2-week expression time which allows for an additional 2–3 cell population doubling before confluency is reestablished. Although not observed in level I dishes, type III foci were observed in level II dishes following exposure to 6-AC (table III). Specifi-

*Table II.* Simultaneous determination of cytotoxicity, mutagenicity and transforming activity of a selected gasoline exhaust extract in the absence of exogenously supplied metabolic activation

| Dose, μg/ml[1] | Cytotoxicity, %[2] | Morphological transformation | | | Gene mutation | |
|---|---|---|---|---|---|---|
| | | T-II | T-III | T-III/$10^4$ survivors | $OUA^R$ | $OUA^R/10^6$ survivors |
| 0 | 0 | 0 | 0 | < 0.24 | 0 | < 1.1 |
| 3 | 11 | 0 | 2 | 0.59 | 7 | 7.7 |
| 10 | 0 | 3 | 3 | 0.71 | 11 | 11.4 |
| 25 | 6 | 1 | 1 | 0.27 | 4 | 5.7 |
| 50 | 20 | 2 | 2 | 0.59 | 0 | < 1.3 |
| 100 | 11 | 0 | 1 | 0.27 | 0 | < 1.0 |
| 250 | 3 | 1 | 2 | 0.49 | 1 | 1.1 |
| 500 | 0 | 0 | 0 | < 0.23 | 4 | 4.8 |

T-II = Number of morphologically transformed type II foci per condition (11–12 plates); T-III = number of morphologically transformed type III foci per condition (11–12 plates); T-III/$10^4$ survivors = number of type III foci per $10^4$ colony-forming cells seeded; $OUA^R$ = number of ouabain-resistant mutants per condition (11–12 plates); $OUA^R/10^6$ survivors = number of ouabain-resistant mutants per $10^6$ colony-forming cells exposed to ouabain.

[1] BALB/3T3 C1. A31-1 cells were treated in suspension for 2 h at 36 °C with solvent control or test article concentrations. The cells were subsequently assayed for cloning efficiency (250 cells/plate), transformation ($1 \times 10^4$ cells/plate), and mutagenesis ($2 \times 10^5$ cells/plate).

[2] Cytotoxicity is reported as the reduction in colony-forming cells relative to the solvent control.

cally, 60–90% of the original dishes exposed to 6-AC contained cells with the potential of generating type III foci. These results suggested that genotypically altered cells were present in the original experiment; however, they were not recognized as such because of their failure to form morphologically transformed foci until further subculture. As demonstrated with 3-methylcholanthrene (3-MC) and MNNG, the level II assay did not reduce the transforming activity found in the level I assay. While the level II assay may not be required for routine screening programs, these observations show that inconsistent results in the simultaneous mutation or transformation assays require further studies. This capability will certainly limit the possibility of false-negative results.

The experience of our laboratory with the simultaneous mutation and transformation assay using BALB/3T3 Clone A31 cells is summarized in

*Table III.* Detection of ouabain-resistant mutants and morphologically transformed BALB/3T3 cells by the standard plate assay and the amplification assay

| Treatment[1] | Dose, μg/ml | Gene mutation $OUA^R$/TD[2] | Level I transformation DWF/TD[3] | Level II transformation GWF/TD[4] |
|---|---|---|---|---|
| 6-AC | 0 | 0/12 | 0/12 | 0/12 |
| | 2.5 | 11/12 | 0/12 | 7/12 |
| | 5.0 | 16/24 | 0/24 | 20/24 |
| | 10.0 | 7/12 | 0/12 | 9/12 |
| MNNG | 0.5 | NT | 9/12 | 8/12 |
| | 0.5 | NT | 12/12 | 12/12 |
| 3-MC | 1.0 | NT | 6/12 | 9/12 |

NT = not tested.

[1] $1 \times 10^4$ cells were treated for 24 h at 37 °C with test article concentrations of solvent alone after which (a) level I cells were maintained for 4 weeks, then fixed and stained, and (b) level II cells were maintained for 2 weeks then subcultured into 5 replicate dishes containing $2 \times 10^5$ cells/dish which were maintained for 2 additional weeks, then fixed and stained.

[2] Number of ouabain-resistant mutants per total dishes treated.

[3] Number of dishes with type III foci per total dishes treated.

[4] Number of groups of level II dishes with foci (each 5-dish group derived from a given level I dish) per total groups of level II dishes.

figure 1. While these data demonstrate a good correlation between the two assays (approximately 70%), they also point to an obvious limitation with standardized short-term protocols for use in screening unknown test articles. Standardized protocols fail to allow for chemically induced alterations in cell kinetics which in turn could affect expression time required for the manifestation of a mutation or a transformation event and thereby yield false-negative results, such as with 6-AC. Captan, a pesticide, was also found negative in the standard transformation assay and positive in the level II assay. However, since standardized protocols must be used in order to allow adequate validation and data comparison, it is hoped that examination of multiple genetic endpoints will reduce the incidence of incorrect conclusions based on the results of a single assay.

Examination of a number of genetic endpoints allows for a more extensive analysis of chemically induced damage, especially over the limited dose ranges used in screening programs, and improves conditions for detection of weak mutagens. Multiple endpoints also provide a means for studying the molecular mechanisms associated with genotoxicity. The ability to evaluate

| | | Cell transformation | |
|---|---|---|---|
| | | Positive | Negative |
| $OUA^R$ gene mutation | Positive | Benzo(a)pyrene<br>Aflatoxin $B_1$<br>2-Acetylaminofluorene<br>3-Methylcholanthrene<br>6-Aminochrysene*<br>Captan*<br>Diesel emission extracts<br>Gasoline combustion extracts<br>Coke oven emission extracts<br>Coded compounds–4 | Benzo(e)pyrene<br>Aflatoxin $G_2$<br>4-Acetylaminofluorene<br>Roofing tar extract<br>Coded compounds–9 |
| | Negative | Coded compounds–2 | Acetone<br>Dimethylsulfoxide<br>Ethanol<br>Distilled water<br>Folpet<br>Coded compounds–11 |

* Positive in the level II transformation assay while negative in the standard level I assay

*Fig. 1.* Comparison of simultaneous gene mutation and cell transformation in BALB/c 3T3 clone A31-1 cells.

cytotoxicity, mutagenicity, and neoplastic transformation simultaneously in the same cell line greatly enhances the utility of the test target system, increases confidence in the data, and limits the possibility of false-negative results.

## *Acknowledgement*

Portions of the research reported herein were supported in part by contract 68-02-2271 from the United States Environmental Protection Agency. We thank Ms. *Jill Bradshaw* and Ms. *Robyn East* for their expert assistance in manuscript preparation.

## *References*

1 Brusick, D.; Matheson, D.; Jagannath, D.R.; Goode, S.; Lebowitz, H.; Reed, M.; Roy, G.; Benson, S.: A comparison of the genotoxic properties of tris(2,3-dibrompropyl)-phosphate and tris(1,3-dichloro-2-propyl)phosphate in a battery of short-term bioassays. J. envir. Path. Toxicol. *3:* 207–226 (1980).

2 Carver, J.H.; Salazar, E.P.; Knize, M.G.; Wandrer, D.L.: Mutation induction at multiple gene loci in Chinese hamster ovary cells. The genetic activity of 15 coded carcinogens and noncarcinogens; in Deserres, Ashby, Evaluation of short-term tests for carcinogens, vol. 1, pp. 594–601 (Elsevier/North-Holland, New York 1981).

3 Connell, J.R.: The relationship between sister chromatid exchange, chromosome aberration, and gene mutation induction by several reactive polycyclic hydrocarbon metabolites in cultured mammalian cells. Int. J. Cancer *24:* 485–489 (1979).

4 Curren, R.D.; Kouri, R.E.; Kim, C.M.; Schechtman, L.M.: Mutagenic and carcinogenic potency of extracts from diesel related environmental emissions. Simultaneous morphological transformation and mutagenesis in BALB/c 3T3 cells; in Pepelko, Danner, Clarke, Health effects of diesel engine emissions: Proceedings of an International Symposium, vol. 2, pp. 861–873 (USEPA, Cincinnati 1980).

5 DiPaolo, J.A.; Takano, K.; Popescu, N.C.: Quantitation of chemically induced neoplastic transformation of Balb/3T3 cloned cell lines. Cancer Res. *32:* 2686–2695 (1972).

6 Kakunaga, T.: A quantitative system for assay of malignant transformation by chemical carcinogens using a clone derived from BALB/3T3. Int. J. Cancer *12:* 463–473 (1973).

7 Reznikoff, C.A.; Brankow, D.W.; Heidelberger, C.: Establishment and characterization of a cloned line of C3H mouse embryo cells sensitive to post-confluence inhibition of cell division. Cancer Res. *33:* 3231–3238 (1973).

8 Reznikoff, C.A.; Bertram, J.S.; Brankow, D.W.; Heidelberger, C.: Quantitative and qualitative studies of chemical transformation of cloned C3H mouse embryo cells sensitive to post-confluence inhibition of cell division. Cancer Res. *33:* 3239–3249 (1973).

9 Schechtman, L.M.; Kouri, R.E.: Control of benzo(a)pyrene induced mammalian cell cytotoxicity, mutagenesis, and transformation by exogenous enzyme fractions; in Scott, Bridges, Sobels, Developments in toxicology and environmental science, vol. 2: Progress in genetic toxicology, pp. 307–316 (Elsevier/North-Holland/Biomedical Press, New York 1977).

Donald L. Putman, PhD, Division of Toxicology/Oncology, Microbiological Associates, 5221 River Road, Bethesda, MD 20816 (USA)

Safety Evaluation and Regulation of Chemicals. 1st Int. Conf., Boston, Mass., 1982, pp. 142–150 (Karger, Basel 1983)

# Cell-Mediated Mutagenesis, an in vitro Approach to Study Organ Specificity of Chemical Carcinogens

*Robert Langenbach, Stephen Nesnow*

Carcinogenesis and Metabolism Branch, Genetic Toxicology Division, Health Effects Research Laboratory, US Environmental Protection Agency, Research Triangle Park, N.C., USA

The phenomena of species, organ, sex and age specificity in response to chemical carcinogens pose an intriguing problem and may be important for the extrapolation of in vivo and in vitro carcinogenesis data to humans. Many carcinogens exist in nonreactive form and must be metabolically activated to manifest their biological activity [8, 21]. We have been interested in developing in vitro systems in which the organ and species specificity of chemical carcinogens can be studied [14–18, 29].

Various approaches for clarifying the causes of organ and species specificity have been employed [1, 2, 5, 8–11, 24]. In the earlier studies, homogenates of various tissues were used for metabolic activation of chemicals [9–11, 24]; however, metabolite profiles, DNA-adduct profiles, and induced mutagenic responses of cell homogenates differ from those of intact cells [3, 4, 6, 23]. More recent studies have looked at the nature of DNA adducts formed in cells or perfused tissues [2, 9, 11] or at the nature and persistence of DNA adducts in living tissues [1, 5]. Differences in the results from these various approaches are apparent. The ability to measure the biological response of a system as well as parameters such as chemical metabolism, DNA adducts, DNA repair, etc., provides a useful system for clarifying some of the biological mechanisms leading to the phenomenon of specificity.

Intact cells are obtained from adult male rat liver, kidney, lung and bladder by enzymatic dispersion, and the cells are cocultivated with Chinese hamster V79 cells in the presence of the test chemical [14–18, 29]. The mutation of the V79 cells from ouabain or 6-thioguanine (TG) sensitivity to resistance then indicates the production of mutagenic/carcinogenic metabolites of the chemical. The V79 cells alone cannot activate the chemicals being studied; therefore, the production of mutagenic intermediates reflects the specificities of activation of the different primary cell types. Factors other than metabolism, believed to be important in determining organ or species specificities [19] (such as DNA repair, nature of DNA adducts, cell uptake of the chemical, etc.), can also be studied and correlated with the mutagenesis end point using this approach.

In the present report, carcinogens from three classes of chemicals were studied for mutagenic activity with primary cells from four organs. The metabolism of the carcinogen benzo(a)pyrene (BP) was studied in detail in all cell systems by high pressure liquid chromatography (HPLC). An attempt was made to correlate metabolism with mutagenic activities and also to correlate cell-mediated mutagenesis data with in vivo carcinogenesis data.

## *Materials and Methods*

Male Sprague-Dawley rats, 6–10 weeks old, were used to obtain the primary liver, lung, kidney and bladder cells. Methods for preparing liver, lung and bladder cells were reported previously [14, 15, 17, 18]. Kidney cells were dissociated enzymatically for 1.5 h at 37 °C from minced kidney tissue with 0.13% trypsin and 0.01% EDTA containing 50 units/ml collagenase. All primary cell types were more than 90% viable at the time of seeding. The medium used was Williams Medium E plus 10% heat-inactivated fetal bovine serum supplemented with 2 m*M* *L*-glutamine, penicillin (100 units/ml) and streptomycin (100 μg/ml). Humidified incubators maintained the cells at 37 °C with an atmosphere of 5% $CO_2$ in air.

Mutation protocols were reported previously [14, 15, 17, 18]. The background mutation frequency was 1 mutant/$10^6$ survivors for ouabain and $\leqslant 3$ mutants/$10^6$ survivors for TG. The expression time was 2 days for ouabain resistance and 6 days for TG [16, 18].

The BP metabolism studies were carried out as previously reported [18].

## *Results*

The basic protocols for cell-mediated mutagenesis studies were reported previously [14–18, 29]. In all the systems, the ratio of activating cells to V79 cells is about 3:1 or 4:1. Using the hepatocyte system as an example, the hepatocytes attach on top of the V79 cells, forming a near monolayer. The medium is then changed to fresh medium containing the test chemical, and the cocultivation is continued for 48 h. The cells are then reseeded to determine toxicity and mutation to the V79 cells. The activating cells do not reattach and grow in this procedure. In this system, the liver cell metabolizes the chem-

*Table I.* Mutagenic activities of BP and $AFB_1$ in the hepatocyte- and embryonic fibroblast-mediated mutagenesis system

| Activating cell | Chemical | Mutants/$10^6$ survivors |
|---|---|---|
| None | – | 1 |
| Liver | BP | 2 |
| Liver | $AFB_1$ | 39 |
| Fibroblasts | BP | 96 |
| Fibroblasts | $AFB_1$ | 2 |

ical to a reactive intermediate that is then transported presumably via cell-to-cell junctions to the V79 cells which are mutated. Proximity between the activating cell and the target cells may be necessary [13] for mutagenesis to occur, although the more stable the intermediate or reactive form, the greater the distance it can transverse.

The hepatocyte to V79 cell arrangement may be an in vitro model for the events that occur in vivo during the initiation of transformation or mutation. In other words, the differentiated cell (hepatocyte, for example) has the metabolic capabilities to carry out the tissues' metabolic functions, but lacks or has limited replicative capabilities. The less well-differentiated cells, including stem cells (represented by the V79 cells) that possibly have lower metabolic capabilities but can replicate, are arranged in proximity to the differentiated cells. Therefore, in vivo the terminally differentiated cells metabolize the xenobiotic, and the activated intermediates are transported to the stem cells. The mutated or transformed stem cells on replication could give rise to the mutant colony or tumor. Such a phenomenon could also explain the long-known difficulty in obtaining mutations or transforming differentiated cells from adult animal tissues in vitro.

Table I summarizes initial attempts to study the cell type specificity in the activation of chemical carcinogens [15]. BP, a potent skin and lung carcinogen capable of also producing fibrosarcomas, and aflatoxin $B_1$ ($AFB_1$), a potent liver carcinogen, were studied for mutagenic activity in the rat embryonic fibroblast- and the adult rat hepatocyte-mediated mutagenesis systems. BP was mutagenic in the embryonic fibroblast-mediated system, while in the hepatocyte-mediated system no significant increase in mutagenic activity was observed with up to 3 $\mu g/ml$ BP. In contrast, $AFB_1$ was not mutagenic in the fibroblast-mediated system but was mutagenic in the hepatocyte-mediated system. This showed that a cell type specificity could be observed

*Table II.* Relative mutagenic activities in the rat cell-mediated mutagenesis systems

| Chemical | Activating cell type | | | |
|---|---|---|---|---|
| | liver | lung | kidney | bladder |
| Anthracene, 3 μg/ml | 1 | 1 | 1 | 1 |
| BP, 1 μg/ml | 2 | 56 | 5 | 6 |
| DMBA, 1 μg/ml | 14 | 154 | 12 | 21 |
| DMN, 100 μg/ml | 95 | 2 | 19 | 11 |
| $AFB_1$, 3 μg/ml | 42 | 4 | ND | ND |

ND = Not determined.

in vitro, which agreed with the in vivo tumorigenic response of the two cell types.

We were concerned about two aspects of this initial study. First, we were comparing mixed embryonic cells to cells from an adult animal organ. Secondly, fibroblasts were being compared to epithelial cells for activation potential, while the majority of tumors in man and animals arise in the epithelium of various organs. In an attempt to overcome the limitations of our initial approach, we began to develop a method for conducting cell-mediated mutagenesis studies with epithelial cells from various adult tissues [16–18, 29]. Table II shows the relative mutagenic activity of four potent chemical carcinogens in the liver, lung, bladder and kidney cell-mediated mutagenesis systems. All of these studies used primary cultures of intact cells from the respective organs and the cultures were predominantly of epithelial-like cells. As can be seen in table II, the noncarcinogenic hydrocarbon anthracene was not mutagenic in any of the cell-mediated systems. In the liver cell-mediated system, BP was inactive as indicated above, while 7,12-dimethylbenz(a)-anthracene (DMBA) was somewhat mutagenic, and dimethylnitrosamine (DMN) was highly mutagenic in this system. By contrast in the lung cell-mediated system, the two hydrocarbons were mutagenic, while DMN was inactive. The kidney cells were able to activate the two hydrocarbons and the nitrosamine. Likewise, cells from the bladder epithelium activated all three carcinogens. $AFB_1$ was active in the liver system but only weakly active in the lung system [18]. The data in table II are expressed as the increase over background per $1.5 \times 10^6$ activating cells. For the liver and lung cell systems, ouabain, which gave a background of 1 mutant/$10^6$ surviving V79 cells [18], was used as the selective agent. However, for the kidney and bladder cell-

*Table III.* BP-diol metabolites formed by liver, lung, bladder, and kidney cells

| Cell type | pmol formed/48 h/$10^6$ cells | | |
|---|---|---|---|
| | pre-9,10-diol | 4,5-diol | 7,8-diol |
| Lung | 4,870 | 90 | 750 |
| Liver | 2,320 | 217 | 212 |
| Bladder | 2,655 | 108 | 415 |
| Kidney | 765 | 50 | 115 |

mediated systems [16], TG was used as the selective agent, and the spontaneous frequency was $3 \times 10^6$ mutant/$10^6$ surviving V79 cells. We reported previously that by applying this method for certain chemicals, the relative activities can be compared even when different selective agents are used [6]. The generality of this approach, however, has yet to be established.

After determining the mutagenic activities of the carcinogens in the various cell-mediated systems, we looked at carcinogen metabolism to ascertain whether correlations between metabolism and mutagenesis could be discerned. Previous studies reported the HPLC profiles of BP metabolism for primary cells [16–18]. All cell lines produce metabolites that are organically soluble or are conjugated to glucuronic acid or sulfate [16, 18]. Table III shows the amounts of three types of metabolites produced by the primary cells in culture. These particular metabolites were chosen for discussion here because of their potential involvement in the transformation or mutation process. The pre-9,10-diols constitute the early eluting fraction from the HPLC and presumably contain tetrols, triols, and other oxygenated metabolites, some of which, probably at one stage in their metabolism, were intermediates that could have reacted with cellular macromolecules. The BP-4,5-diol probably results from the BP-4,5-oxide, a species also capable of interacting with cellular constituents, although the 4,5-diol itself is weakly active as a mutagen for V79 cells [6]. The BP-7,8-diol is a precursor to the BP-7,8-diol-9,10-epoxide, the presumed active known metabolite of BP [30]. As can be seen in table III, the order of pre-9,10-diol production is lung > liver > bladder > kidney, which also is the order of total BP metabolism, including conjugates. The liver produces the greatest amount of the BP-4,5-diol. The BP-7,8-diol, thought to be a precursor to the most important biological metabolite, is formed mostly in the lung, the cell type where BP was also most mutagenic. Bladder and kidney cells, which also activate BP to mutagenic

*Table IV.* Comparison of relative carcinogenic (car.) and mutagenic (mut.) activities[1]

| Chemical | Target organ or activating cell type | | | | | | | |
|---|---|---|---|---|---|---|---|---|
| | liver | | lung | | kidney | | bladder | |
| | car. | mut. | car. | mut. | car. | mut. | car. | mut. |
| BP | ± | − | ++ | ++ | + | + | − | + |
| DMBA | + | + | ++ | ++ | ++ | + | ± | + |
| DMN | ++ | ++ | ± | − | ++ | + | ? | + |
| $AFB_1$ | ++ | ++ | + | ± | + | ND | + | ND |

ND = Not determined.
[1] Mutation: − = < 3 × background; + = 4–25 × background; ++ = > 25 × background.

intermediates, produce significant levels of the 7,8-diol. However, liver cells, which do not activate BP to mutagenic intermediates, also produced the 7,8-diol. Therefore, the absolute amount of the BP-7,8-diol produced does not directly correlate with the mutagenic reponse in the cell-mediated systems.

## *Discussion*

The data reported in this paper suggest that the cell-mediated mutagenesis approach may be useful for the study of some aspects of the organ specificity of chemical carcinogens. In table IV we compared the relative carcinogenic activities of the chemicals to their relative mutagenic activities in the cell-mediated system. As can be seen in table IV, there is general agreement in the activities between the in vivo and in vitro results. The relative carcinogenic activities for liver, lung, and kidney were assigned from literature references [19, 20, 22, 25]; rodent bladder carcinogenesis data with these chemicals are limited. DMBA has been reported to cause focal hyperplasia and papillomas in the bladder [25] and to transform bladder epithelial cells in organ culture [28]. The finding that bladder epithelial cells can activate DMN is important, given its presence in urine of humans with certain types of urinary tract infections [26]. While DMN has not been shown to induce bladder cancer in animals or humans, people with frequent urinary tract infections have an increased incidence of bladder cancer [7]. BP also has not been shown to induce bladder cancer, although correlations between bladder cancer and smoking have been shown [27], and BP, along with other known carcinogens, is a component of cigarette smoke. The possible involvement of BP glucuronides in bladder cancer has been suggested [12]. The present studies predict that these carcinogens may induce bladder cancer and demonstrate the role of the bladder epithelium in the activation of some chemical carcinogens.

The study of BP metabolism in primary cells was undertaken because of the differential mutagenic response to it in the various organs. All the cell types are capable of metabolizing BP, although quantitative differences in the total metabolism and individual metabolites existed. Glucuronidation and formation of sulfates of BP also occurred in all tissues. However, as stated above, a precise correlation between any metabolite or pattern of metabolites and mutagenic activity could not be discerned. Again, this result emphasizes that metabolism is one of many factors leading to organ and species specificity.

The use of human cells from various tissues for metabolic activation and mutable or transformable human cells as the target cells would be a worthwhile future application of the cell-mediated approach. In this in vitro all-human cell system, one would be studying directly the effect of the chemical being applied to human tissues. Concomitant with this approach, the other biological parameters involved (see Introduction) could be measured. This battery of biological data then could be used to determine the target organ and potential risk of the agent to man. The accumulation of such data may provide a method for extrapolating risk to humans that would not involve extrapolation of risk from in vivo rodent studies. Such an approach appeals, because it does not depend on the long-term maintenance of differentiated human cells in culture, a difficult task to date. Thus, the potential use of the cell-mediated approach appears to have fundamental as well as applied uses in the study of chemical carcinogenesis.

## *References*

1 Autrup, H.; Wefald, F.C.; Jeffrey, A.M.; Tate, H.; Schwartz, R.D.; Trump, B.F.; Harris, C.C.: Metabolism of benzo(a)pyrene by cultured tracheobronchial tissues from mice, rats, hamsters, bovines and humans. Int. J. Cancer *25:* 293–300 (1980).

2 Bartsch, H.; Malaveille, E.; Montesano, R.: In vitro metabolism and microsome-mediated mutagenicity of dialkylnitrosamines in rat, hamster, and mouse tissues. Cancer Res. *35:* 644–651 (1975).

3 Bigger, C.A.H.; Thomaszewski, J.E.; Dipple, A.: Differences between products of binding of 7,12-dimethylbenz(a)anthracene to DNA in mouse skin and in a rat liver microsomal system. Biochem. biophys. Res. Commun. *80:* 229–235 (1978).

4 Bigger, C.A.H.; Tomaszewski, J.E.; Dipple, A.; Lake, R.S.: Limitations of metabolic activation systems used with in vitro tests for carcinogens. Science *209:* 503–505 (1980).
5 Boroujerdi, M.; Kung, H.; Wilson, A.G.E.; Anderson, M.W.: Metabolism and DNA binding of benzo(a)pyrene in vivo in the rat. Cancer Res. *41:* 951–957 (1981).
6 Bradley, M.O.; Bhuyan, B.; Francis, M.C.; Langenbach, R.; Peterson, A.; Huberman, E.: Mutagenesis by chemical agents in V79 Chinese hamster cells: a review and analysis of the literature. Mutat. Res. *87:* 81–142 (1981).
7 Cox, C.E.; Cass, A.S.; Boyce, W.H.: Bladder cancer: a 26-year review. J. Urol. *101:* 550–565 (1969).
8 Heidelberger, C.: Chemical carcinogenesis. A. Rev. Biochem. *44:* 79–121 (1975).
9 Hundley, S.G.; Freudenthal, R.E.: A comparison of benzo(a)pyrene metabolism by liver and lung microsomal enzymes from 3-methyl-cholanthrene-treated Rhesus monkeys and rats. Cancer Res. *37:* 3120–3125 (1979).
10 Hutton, J.J.; Meier, J.; Hackney, C.: Comparison of the in vitro mutagenicity and metabolism of dimethylnitrosamine and benzo(a)pyrene in tissues from inbred mice treated with phenobarbital, 3-methylcholanthrene or polychlorinated biphenyls. Mutat. Res. *66:* 75–94 (1979).
11 Juchau, M.R.; DiGiovanni, J.; Namkung, M.J.; Jones, A.H.: A comparison of the capacity of fetal and adult liver, lung, and brain to convert polycyclic aromatic hydrocarbons to mutagenic and cytotoxic metabolites in mice and rats. Toxicol. appl. Pharmacol. *49:* 171–178 (1979).
12 Kinoshita, N.; Gelboin, H.V.: $\beta$-Glucuronidase catalyzed hydrolysis of benzo(a)pyrene-3-glucuronide and binding to DNA. Science *199:* 307–309 (1978).
13 Kuroki, T.; Drevon, C.: Direct or proximate contact between cells and metabolic activation systems is required for mutagenesis. Nature, Lond. *271:* 368–370 (1978).
14 Langenbach, R.; Freed, H.J.; Huberman, E.: Liver cell-mediated mutagenesis of mammalian cells by liver carcinogens. Proc. natn. Acad. Sci. USA *75:* 2864–2867 (1978).
15 Langenbach, R.; Freed, H.J.; Raveh, D.; Huberman, E.: Cell specificity in metabolic activation of aflatoxin $B_1$ and benzo(a)pyrene to mutagens for mammalian cells. Nature, Lond. *276:* 277–280 (1978).
16 Langenbach, R.; Malick, L.; Nesnow, S.: Bladder cell-mediated mutagenesis. J. natn. Cancer Inst. *66:* 913–917 (1981).
17 Langenbach, R.; Nesnow, S.; Malick, L.; Gingell, R.; Tompa, A.; Kuszynski, C.; Leavitt, S.; Sasseville, K.; Hyatt, B.; Cudak, C.; Montgomery, L.: Organ specific activation of carcinogenic polynuclear aromatic hydrocarbons in cell culture; in Cooke, Dennis, Chemical analysis and biological fate: polynuclear aromatic hydrocarbons, pp. 75–84 (Battelle Press, Columbus 1981).
18 Langenbach, R.; Nesnow, S.; Tompa, A.; Gingell, R.; Kuszynski, C.: Lung and liver cell-mediated mutagenesis systems: specificities in the activation of chemical carcinogens. Carcinogenesis *2:* 851–858 (1981).
19 Magee, P.N.; Montesano, R.; Preussman, R.: N-Nitroso compounds and related compounds; in Searle, Chemical carcinogens. Am. Chem. Soc. Monogr. 173, pp. 491–625 (American Chemical Society, Washington 1976).
20 Marquardt, H.: Carcinogenic activity of DMBA in the liver. Cancer Res. *33:* 1102–1109 (1973).
21 Miller, J.A.: Carcinogenesis by chemicals: an overview. Cancer Res. *30:* 559–576 (1970).

22 Nettesheim, P.; Griesemer, R.A.: Experimental models for studies of respiratory tract carcinogenesis; in Harris, Pathogenesis and therapy of lung cancer, pp. 75–188 (Dekker, New York 1978).

23 Newbold, R.F.; Wigley, C.B.; Thompson, M.H.; Brookes, P.: Cell-mediated mutagenesis in cultured Chinese hamster cells by carcinogenic polycyclic hydrocarbons: nature and extent of the associated hydrocarbon-DNA reaction. Mutat. Res. *43:* 101–116 (1977).

24 Prough, R.A.; Patrizi, V.W.; Okita, R.T.; Masters, B.S.S.; Jakobsson, S.W.: Characteristics of benzo(a)pyrene metabolism by kidney, liver, and lung microsomal fractions from rodents and humans. Cancer Res. *39:* 1199–1206 (1979).

25 Public Health Service, National Institutes of Health: Survey of compounds which have been tested for carcinogenic activity; DHEW publication No. (NIH) 73–75, PHS publication No. 149 (US Government Printing Office, Washington 1961–1967).

26 Radomski, J.L.; Greenwald, D.; Hearn, W.L.; Block, H.L.; Woods, F.W.: Nitrosamine formation in bladder infections and its role in the etiology of bladder cancer. J. Urol. *120:* 48–50 (1978).

27 Stevens, R.G.; Moolgavkor, S.H.: Estimation of relative risk from vital data: smoking and cancers of the lung and bladder. J. natn. Cancer Inst. *63:* 1351–1357 (1979).

28 Summeshayes, I.C.; Franks, L.M.: Effects of donor age on neoplastic transformation of adult mouse bladder epithelium in vitro. J. natn. Cancer Inst. *62:* 1017–1023 (1979).

29 Tompa, A.; Langenbach, R.: Culture of adult rat lung cells: benzo(a)pyrene metabolism and mutagenesis. In Vitro *15:* 569–578 (1979).

30 Yang, S.K.; Deutsch, J.; Gelboin, H.V.: Benzo(a)pyrene metabolism: activation and detoxification; in Gelboin, Ts'o, Polycyclic hydrocarbons and cancer, pp. 205–224 (Academic Press, New York 1978).


Dr. R. Langenbach, Genetic Toxicology Division, Environmental Protection Agency, Research Triangle Park, NC 27711 (USA)

Safety Evaluation and Regulation of Chemicals. 1st Int. Conf., Boston, Mass., 1982, pp. 151–159 (Karger, Basel 1983)

# The Use of Human Tissues and Cells in Carcinogen Metabolism and Toxicity Studies

*H. Autrup, J.F. Lechner, C.C. Harris*[1]

Laboratory of Human Carcinogenesis, National Cancer Institute, Bethesda, Md., USA

One of the difficulties in biomedical research is to relate observations made in experimental animals to the human situation, and, therefore, to the identification of potential toxic and/or carcinogenic risk to humans. Experimentation with human tissues is complicated because of logistic and ethical reasons. Currently, most studies in normal human tissues use either skin fibroblast or subcellular fractions prepared from surgical or autopsy tissues. During the last decade, investigators in several laboratories have successfully cultured the major types of human epithelial tissues both as explants and as cell cultures, and have opened up a new area of toxicology and cancer research. The development of in vitro models of target tissues, furthermore, allows one to do comparative studies in humans and experimental animals at the tissue, cellular, and subcellular levels of biological organization. This advance is aiding efforts to extrapolate in vivo observations and mechanisms from experimental animals to human.

Human tissues and cells have been used both as the activating system and as the detector system in assays measuring damage to DNA, such as mutagenesis, unscheduled DNA synthesis and sister chromatid exchanges [1, 8, 20, 23, 28]. Several studies on mutagenesis and malignant transformation of human skin fibroblasts have been reported [21, 24, 27, 31]. One of the disadvantages of using established fibroblast cell lines for this type of study is

[1] We are indebted to the collaborative efforts of our co-workers including *J. Essigman, A. Haugen, A. Jeffreys, G. Stoner,* and *B. Trump*. The secretarial aid of *N. Paige* is appreciated.

the negligible activity of mixed function oxidase in the cells, an enzyme complex responsible for the activation of many xenobiotics. However, primary fibroblast cultures and cells at early passage number still have the ability to metabolize many potential chemical carcinogens although to a lesser extent than the epithelial cells from the same tissue [2, 7, 30]. Another disadvantage is that fibroblasts are rarely the target cells in vivo during carcinogenesis or organ toxicity.

We will review here (a) the culture conditions for human epithelial tissues and cells; (b) the ability of these tissues and cells to activate chemical carinogens as measured by metabolite profile, binding of metabolites to cellular macromolecules, and carcinogen-DNA adducts, and (c) cytotoxicity of various chemical and physical agents to human epithelial and fibroblastic cells as measured by colony-forming efficiency.

## *Material and Methods*

### *Culture of Human Epithelial Tissues*

Nontumorous specimens from adult donors with or without cancer are collected at the time of surgery and autopsy [17]. Although tissues from 'immediate autopsy' are preferred, several laboratories have successfully initiated cultures from some tissues obtained by routine autopsy 2–6 h after death of the patients [17]. Specimens are transported to the laboratory in L-15 culture medium at 4°C. In our laboratory, methods have been established to culture bronchus, colon, pancreactic ducts, and esophagus as explants in a chemically-defined medium [17]. Primary bronchial epithelial cell cultures have been initiated from outgrowth of bronchus explants [26]; by serially transferring these explants several cultures can be initiated. Outgrowths of epithelial cells have also been established from bronchial explants cryopreserved for up to 6 months [33]. We and our co-workers have successfully cultured bronchial epithelial cells in a serum-free medium [26]. Explants of human bronchus can be maintained for at least 12 months, while explants of other organs have been reported viable for up to 4 months.

### *Carcinogen Metabolism*

After, generally, 7 days in culture to minimize the effects of exogenous agents, such as drugs and dietary constituents, the explants were incubated with radioactively labelled carcinogens for 24 h. DNA was isolated from the surface epithelia scraped from the explants, and used for both quantitation and determination of the carcinogen-DNA interaction [6]. Metabolites released into the media were extracted and analyzed by chromatographic procedures [6].

### *Cytotoxicity Assay*

Clonal growth rate was used to measure cytotoxicity [18]. After 24 h, the medium was replaced with medium containing the compound under investigation, and the cell was exposed for 1–3 days, 10 days postincubation, the cells were fixed in formalin and numbers of colonies and the number of cells in colonies were determined.

*Table I.* Binding level of BP and AFB to DNA in cultured human tissues

| Tissue | Culture | BP | AFB | Interindividual variation in BP | Reference |
|---|---|---|---|---|---|
| Bladder | E | 6.4 | 1.7 | 55 | 34 |
| Bronchus | E | 1.0 | 0.5 | 75 | 2, 3 |
| Colon | E | 0.3 | 0.1 | 150 | 2, 3 |
| Duodenum | E | 0.6 | 0.4 | 50 | 2, 3 |
| Endometrium | E | 0.1 | n.t. | 70 | 11 |
| Esophagus | E | 0.7 | 0.4 | 99 | 2, 3 |
| Keratinocytes | C | 2.9 | n.t. | – | 23, 35 |
| Liver (fetal) | E | n.t. | 0.7 | – | 2 |

The results are expressed as number of modification per $10^6$ nucleotides. Concentration of carcinogens used was 1–1.5 $\mu M$, incubation time was 24 h.
E = explants; C = cells; n.t. = not tested.

## *Results and Discussion*

### *Carcinogen-DNA Interaction*

Human epithelial tissues have the ability to activate several chemical classes of procarcinogens into metabolites that bind to cellular DNA. Generally, a higher level of binding was seen in cell cultures than in explant culture. This could, in part, be due to the fact that cells with higher metabolic activity might have been selected whereas the explants represent a complex mixture of different cell types. While all the organs metabolized all PAHs so far tested, some organ specificity within the sensitivity limits of this methodology, was observed with N-nitrosamines. The lack of detectable binding in some tissues could be due to the low specific activity of the radiolabelled N-nitrosamines. The highest mean level of binding to DNA of both BP and AFB was seen in bladder explants (table I). When compared to tissues from experimental animals, the binding values of PAH to DNA was generally higher in human tissues [4–6, 34]. Wide quantitative differences in the binding values of BP (table I) and other carcinogens [15] have been found in cultured human tissues. The magnitude of these differences ranges from 50- to 150-fold and is similar to that found in pharmacogenetic studies [15]. However, this variation in human tissues makes a quantitative correlation of animal data to the human situation more complicated.

*Table II.* Metabolism of B(a)P to organo-soluble metabolites by human and rat tissues

| Metabolite | Rat | | | Human | | |
|---|---|---|---|---|---|---|
| | colon | bladder | bronchus | colon | bladder | bronchus |
| Tetrols | 15 | 20 | 13 | 4 | 19 | 21 |
| 9,10-Diol | 20 | 20 | 12 | 19 | 21 | 28 |
| 7,8-Diol | 8 | 13 | 6 | 9 | 10 | 5 |
| 9-Hydroxy | 4 | 2 | 8 | 2 | 3 | 4 |
| 3-Hydroxy | 4 | 10 | 8 | 27 | 10 | 2 |

After incubation of the explant with [$^3$H] BP for 24 h, the media were extracted with ethylacetate/acetone, and the metabolites were separated by high pressure liquid chromatography.

The adducts formed in vitro between the carcinogen metabolites and DNA are quite similar in both different human organs and tissues from experimental animals in which the compounds are carcinogenic [5, 6]. The carcinogen-DNA adducts formed in epithelial cell cultures are similar to those formed in explants [2, 4], but are generally different from the adducts formed in using liver microsomes as the activating system and native DNA as recipient [9]. However, the BP-DNA adducts in tracheobronchial tissues formed in vitro are different from the adducts formed in rat lung in vivo after i.p. injection of BP [10]. This discrepancy is probably due to the fact that most of the BP is metabolized in the liver and not in the target organ. Guanosine was the major target for the reaction between most of the activated carcinogens and DNA. However, different chemical types of carcinogens react with different positions in guanosine. Polycyclic aromatic hydrocarbons, e.g., BP, react predominantly with the exocyclic 2-amino group, while AFB reacts mostly with the 7-position. This position is also the major site for alkylation by N-nitrosamines, although minor reaction products, e.g., alkylation at the 0–6 position may be responsible for the mutagenic-carcinogenic action of this type of compound.

*Carcinogen Metabolism*

In intact tissues and cells, carcinogens, such as BP and AFB, are metabolized into both organo-soluble and water-soluble metabolites. Bronchial tissues are more efficient in converting primary metabolites of BP into sulfate esters, glucuronides and glutathione conjugates than the other organs [3]. This is reflected in a relatively lower amount of BP phenols (table II) that are the major substrates for conjugation reactions. No qualitative differences between the metabolite profile of BP are seen between different human organs and the corresponding rat tissues although quantitative differences do exist. This is probably due to different relative amounts of the cytochrome P-450s in various organs and animal species. Different forms of cytochrome P-450 catalyze the different steps in the activation of BP to its ultimate carcinogenic form [13]. Similarly, the predominant form of the cytochrome P-450 converting AFB into its mutagenic metabolites is identical to the form converting BP to a nonmutagenic metabolite [14]. In our studies with human colon and bronchus, we did not find any correlation between the binding levels of BP and AFB [1] but found a positive correlation between compounds with greater chemical similarities such as BP and 7,12-dimethylbenz(a)anthracene (DMBA) [2]. These results indicate the difficulties in correlating observations from one type of compound to another unless their metabolic pathways are clearly understood.

*Tissue- and Cell-Mediated Mutagenesis*

Pulmonary alveolar macrophages (PAM) and bronchial explants have been used as the activation system in a cell- or tissue-mediated mutagenesis assay. Both converted BP and BP 7,8-diol into an ultimate mutagen(s) detected in co-cultivated Chinese hamster V-79 cells [16, 19, 20]. Increases in both frequencies of ouabain-resistant mutation and sister chromatid exchanges were found in V-79 cells only when they were co-incubated with procarcinogens and either PAM or bronchial explants; as expected, the proximate carcinogen, BP 7,8-diol caused higher mutation frequency than the parent compound. A 30-fold interindividual variation was also detected using this approach. A positive correlation between the mutation frequency in the V-79 cells, and the binding of BP to DNA in the bronchial explants was noted. N-nitrosodimethylamine did not induce any detectable mutation rates in V-79 cells when bronchial explants were used as the activator system.

*DNA Repair*

DNA single-strand breaks, generated during excision repair, are another measure for the reaction between carcinogens and DNA. Single-strand breaks have been quantitated in human bronchial epithelial and fibroblast cells by the alkaline elution technique [22] after treatment with both X-rays, UV radiation, and direct and indirect acting chemical carcinogens. Treatment of bronchial epithelial cells with either DMNA or amosite did not induce any single strand breaks in the presence of Ara-C and hydroxyurea, two DNA polymerase inhibitors, whereas DMBA, BPDE, and MNNG induced a substantial number of breaks [12]. Repair replication as measured by BND chromatography [11] induced by BPDE is significantly higher in epithelial cells than in fibroblasts.

*Cytotoxicity*

Cytotoxic effect of asbestos and other chemical carcinogens can be estimated by measuring the inhibition of clonal cell growth. 4-Nitroquinoline-N-oxide, a direct-acting carcinogen, is extremely toxic to prostate epithelial cells, while 7, 12-DMBA, an indirect-acting carcinogen that is metabolized by bronchial explants to a significant extent, is very toxic to bronchial epithelial cells. Chrysolite is the most toxic of the asbestos fibers. It is interesting to note that the fibroblastic cells from the same donor are comparatively more resistant to the toxic effect of amosite than the epithelial cells. The clonal assay procedure provides an easy and inexpensive system to assess the effect of toxic chemicals on human target cells not only on plating efficiency, but also on the growth rate of the cells.

*Conclusions*

Methods to culture human epithelial tissues and cells have been developed in order to study the mechanism of carcinogenesis in human target tissues. Studies on the metabolism of various classes of chemical carcinogens indicate that the metabolic pathways – metabolic profile and carcinogen-DNA adducts – are similar in tissues from human and experimental animals in which these compounds are carcinogenic. However, quantitative differences exist among both organs and species. Several types of carcinogens showed cytotoxicity and induced DNA-damage in human bronchial epithelial and fibroblastic cells. These studies into the mechanism of carcinogenesis are still at very early stages of development and more research using cultured human tissues and cells is required before the methods can be used for safety evaluation.

## *References*


1 Aust, A.E.; Falahee, K.J.; Maher, V.M.; McCormick, J.J.: Human cell-mediated benzo(a)pyrene cytotoxicity and mutagenicity in human diploid fibroblast. Cancer Res. *40:* 4070–4075 (1980).

2 Autrup, H.: Metabolism of chemical carcinogens by human tissues. Drug Metab. Rev. *13:* 603–648 (1982).

3 Autrup, H.; Grafstrom, R.C.; Brugh, M.; Lechner, J.F.; Haugen, A.; Trump, B.F.; Harris, C.C.: Comparison of benzo(a)pyrene metabolism in bronchus, colon, esophagus and duodenum from the same individual. Cancer Res. *42:* 934–938 (1982).

4 Autrup, H.; Grafstrom, R.C.; Christensen, B.; Kieler, J.: Metabolism of chemical carcinogens by cultured human and rat bladder epithelial cells. Carcinogenesis *2:* 763–768 (1981).

5 Autrup, H.; Harris, C.C.; Fugaro, S.; Selkirk, J.K.: Effect of various chemicals on the metabolism of benzo(a)pyrene by cultured rat colon. Chem.-Biol. Interact. *18:* 337–347 (1977).

6 Autrup, H.; Wefald, F.C.; Jeffrey, A.M.; Tate, H.; Schwartz, R.D.; Trump, B.F.; Harris, C.C.: Metabolism of benzo(a)pyrene by cultured tracheobronchial tissues from mice, rats, hamsters, bovine and humans. Int. J. Cancer *25:* 293–300 (1980).

7 Bartley, J.; Bartholomew, J.; Stampfer, M.R.: Benzo(a)pyrene metabolism in human mammary epithelial and fibroblast cells. J. supramol. Struct. Cell. Biochem. suppl. 5, p. 165 (1981).

8 Bartsch, H.; Malaveille, C.; Camus, A.-M.; Martel-Planche, G.; Brun, G.; Hautefeuille, A.; Sabadie, N.; Barin, A.; Kuroki, T.; Drevon, C.; Piccoli, C.; Montesano, R.: Validation and comparative studies on 180 chemicals with *S. typhimurium* strains and V79 Chinese hamster cells in the presence of various metabolizing systems. Mutat. Res. *76:* 1–50 (1980).

9 Bigger, C.A.H.; Tomaszewski, J.E.; Dipple, A.: Differences between products of binding of 7,12-dimethylbenz(a)anthracene to DNA in mouse skin and in a rat liver microsomal system. Biochem. biophys. Res. Commun. *80:* 229–235 (1978).

10 Boroujerdi, M.; Kung, H.-C.; Wilson, A.G.E.; Anderson, M.W.: Metabolism and DNA binding of benzo(a)pyrene in vivo in the rat. Cancer Res. *41:* 951–957 (1981).

11 Dorman, B.H.; Genta, V.M.; Mass, M.J.; Kaufman, D.A.: Benzo(a)pyrene binding to DNA in organ cultures of human endometrium. Cancer Res. *41:* 2718–2722 (1981).

12 Fornace, A.J., J.; Grafstrom, R.C.; Lechner, J.F.; Harris, C.C.: Repair of DNA damage in human bronchial epithelial and fibroblastic cells (submitted).

13 Gozukara, E.M.; Belvedere, G.; Robinson, R.C.; Deutsch, J.; Coon, M.J.; Guengerich, F.P.; Gelboin, H.V.: The effect of epoxide hydratase on benzo(a)pyrene diol epoxide hydrolysis and binding to DNA and mixedfunction oxidase proteins. Molec. Pharmacol. *19:* 153–161 (1981).

14 Gurtoo, H.L.; Dahms, R.P.; Kanter, P.; Vaught, J.B.: Association and dissociation of the Ah locus with the metabolism of aflatoxin $B_1$ by mouse liver. Co-segregation of aflatoxin $B_1$ hydroxylase induction with aryl hydrocarbon (benzo[a]pyrene) hydroxylase induction. J. biol. Chem. *253:* 3952–3961 (1978).

15 Harris, C.C.; Grafstrom, R.C.; Trump, B.F.; Autrup, H.: Differences in metabolism of

chemical carcinogens in cultured human epithelial tissues; in Harris, Cerutti, Mechanisms in chemical carcinogenesis, pp. 289–298 (Liss, New York, 1982).

16 Harris, C.C.; HSU, I.-C.; Trump, B.F.; Selkirk, J.K.: Human pulmonary alveolar macrophages metabolize benzo(a)pyrene to proximate and ultimate mutagens. Nature, Lond. *272:* 633–634 (1978).

17 Harris, C.C.; Trump, B.F.; Stoner, G.D.: Methods in cell biology. Normal human tissue and cell culture, vol. 21A, 21B (Academic Press, New York 1980).

18 Haugen, A.; Schafer, P.; Lechner, J.F.; Stoner, G.D.; McDowell, E.F.; Trump, B.F.; Harris, C.C.: Cellular ingestion, toxic effects, and lesions observed in human respiratory epithelium cultured with asbestos fibers. Int. J. Cancer (in press).

19 Hsu, I.-C.; Harris, C.C.; Yamaguchi, M.; Trump, B.F.; Schafer, P.W.: Induction of ouabain-resistant mutation and sister chromatid exchanges in Chinese hamster cells with chemical carcinogens mediated by human pulmonary macrophages. J. clin. Invest. *64:* 1245–1252 (1978).

20 Hsu, I.-C.; Stoner, G.D.; Autrup, H.; Trump, B.F.; Selkirk, J.K.; Harris, C.C.: Human bronchus-mediated mutagenesis of mammalian cells by carcinogenic polynuclear aromatic hydrocarbons. Proc. natn. Acad. Sci. USA *75:* 2003–2007 (1978).

21 Huang, S.L.; Huang, S.-M.S.; Casperson, C.; Waters, M.D.: Induction of 6-thioguanine resistance in synchronized human fibroblast cells treated with methylmethane sulfonate, N-acetoxy-2-acetylaminofluorene and N-methyl-N'-nitro-N-nitrosoguanidine. Mutat. Res. *83:* 251–260 (1981).

22 Kohn, K.W.; Erickson, L.C.; Ewig, R.A.G.; Friedman, C.A.: Fractionation of DNA from mammalian cells by alkaline elution. Biochemistry *15:* 4629–4637 (1976).

23 Kuroki, T.; Menoto, N.; Kitano, Y.: Metabolism of benzo(a)pyrene in human epidermal keratinocytes in culture. Carcinogenesis *1:* 559–565 (1980).

24 Lake, R.S.; Kropko, M.L.; Pezzutti, M.R.; Shoemaker, R.H.; Igel, H.J.: Chemical induction of unscheduled DNA synthesis in human stem epithelial cell cultures. Cancer Res. *38:* 2091–2098 (1978).

25 Lechner, J.F.; Babcock, M.S.; Marnell, M.; Narayan, K.S.; Kaighn, M.E.: Normal human prostate epithelial cell cultures; in Harris, Trump, Stoner, Methods in cell biology, vol. 21B, pp. 195–225 (Academic Press, New York 1980).

26 Lechner, J.F.; Haugen, A.; McClendon, I.A.; Pettis, E.W.: Clonal growth of normal adult human bronchial epithelial cells in a serum-free medium. In Vitro *18:* 633–642 (1982).

27 Milo, G.E.; Dipaolo, J.A.: Presensitization of human cells with extrinsic signals to induce chemical carcinogenesis. Int. J. Cancer *26:* 805–812 (1980).

28 Michalopoulos, G.; Strom, S.C.; Kligerman, A.D.; Irons, G.P.; Novicki, D.L.: Mutagenesis induced by procarcinogens at the hypoxanthine-guanine phosphoribosyl transferase locus of human fibroblasts cocultured with rat hepatocytes. Cancer Res. *41:* 1873–1878 (1981).

29 Moore, B.P.; Hicks, R.M.; Knowles, M.A.; Redgrave, S.: Metabolism and binding of benzo(a)pyrene and 2-acetylaminofluorene by short-term organ cultures of human and rat bladder. Cancer Res. *42:* 642–648 (1982).

30 Parkinson, E.K.; Newbold, R.F.: Benzo(a)pyrene metabolism and DNA adduct formation in serial cultivated strains of human epidermal keratinocytes. Int. J. Cancer *26:* 289–299 (1980).

31 Parsa, I.; Marsh, W.H.; Sutton, A.L.; Butt, K.M.H.: Effects of dimethylnitrosamine on organ cultured adult human pancreas. Am. J. Path. *102:* 403–411 (1981).

32 Scudiero, D.; Henderson, E.; Norin, A.; Strauss, B.: The measurement of chemically-induced DNA repair synthesis in human cells by BND-cellulose chromatography. Mutat. Res. *29:* 473–488 (1975).
33 Stoner, G.D.: Personal commun.
34 Stoner, G.D.; Daniel, F.B.; Schenck, K.M.; Schut, H.A.J.; Goldblatt, P.J.; Sandwisch, D.W.: Metabolism and DNA binding of benzo(a)pyrene in cultured human bladder and bronchus. Carcinogenesis *3:* 195–201 (1982).
35 Theall, G.; Eisinger, M.; Grunberger, D.: Metabolism of benzo(a)pyrene and DNA adduct formation in cultured human epidermal keratinocytes. Carcinogenesis *2:* 581–587 (1981).

Herman Autrup, PhD, Laboratory of Human Carcinogenesis, NCI, NIH, Building 37, Room 2C16, Bethesda, MD 20205 (USA)

Safety Evaluation and Regulation of Chemicals. 1st Int. Conf., Boston, Mass., 1982, pp. 160–165 (Karger, Basel 1983)

# Practical Aspects of Making Industrial Decisions Based on in vitro Data

*L.S. Grosso, B.I. Doerr*

Hoechst-Roussel Pharmaceuticals, Inc., Somerville, N.J., USA

In vitro and in vivo short-term tests are included in safety assessment programs to identify chemicals that have a potential to interact with DNA. This interaction may result in heritable effects if it occurs on germ cells, and in cancer if it occurs on somatic cells. These tests are conducted in response to both regulatory and scientific needs and are useful as guides for decisions regarding the development of experimental compounds. They may also be useful in major decisions on commercial products, provided their data are evaluated along with the results of other toxicological tests in animals as well as a consideration of human exposure and the products' use.

The international requirements for in vivo toxicology studies are becoming more uniform; however, similar agreements for mutagenicity studies have not been achieved. At present, there is no agreement on which studies are to be performed by industry to satisfy various government regulations. Guidelines or recommendations, developed by international groups, for mutagenicity testing include the proposed 1978 EPA Guidelines for Mutagenicity Testing of New Pesticides, and the National Academy of Sciences recommended procedures for evaluating the toxicity of household substances, published in 1977. Others include the OECD Draft Guidelines for health effects testing, from 1980; the European Economic Commission's Committe on Proprietary Medicinal Products, 1980; from Italy, the Guidelines for Evaluating the Mutagenic Effects of Chemicals, 1977, and from England, the Department of Health and Social Security Guidelines for Mutagenicity Testing, 1981.

As an example, the EPA Guidelines for new pesticides are extensive, requiring at least eight assays to test for the three types of mutagenic end points: gene or point mutations, chromosomal aberrations, and primary DNA damage. Tests suitable for detecting gene mutations include assays in bacteria, eukaryotic micro-organisms and mammalian cells in culture and the mouse-specific locus test (any three). Tests for chromosomal aberrations include in vivo cytogenetics in mammals, heritable translocation in rodents, heritable chromosomal effects in insects and dominant lethal effects in rodents (at least three). Primary DNA damage may be assayed in repair tests in bacteria (differential killing), unscheduled DNA synthesis in mammalian cells, mitotic recombination or gene conversion in yeast and Sister chromatid exchange (at least two).

Currently, the EPA allows greater latitude for the types of assays conducted, provided these tests include an evaluation of the three basic endpoints. One of the major objections to the guidelines is the lack of reproducibility of tests conducted in different laboratories. This concern may be alleviated by establishing a basis for comparisons using the results of the EPA's Gene-Tox program.

Also in the US, the FDA has proposed a battery of mutagenicity tests to assess the carcinogenic potential of the uncharacterized components of biomass products. The FDA requested a battery of genetic toxicity tests on polar and nonpolar solvent extractables from the sponsored products and urine from both target and test animals. For point or gene mutation effects, the FDA recommended reversion tests in bacteria and forward mutation assays in cultured mammalian cells or sex-linked recessive lethal tests in *Drosophila*. For DNA damage, the FDA recommended UDS in bacteria or cultured mammalian cells, or the primary rat hepatocyte assay. Industry objected to this proposal and issued the following responses:

– Some of the assays lack validation.

– Sample preparation procedures outlined in the FDA-proposed document could produce artifacts giving false-positive results.

– The proposal was written such that one positive response in one test would indicate a carcinogenic concern of the products' components.

– Lifetime bioassays are more reliable tests for carcinogenicity.

The current thinking at FDA is to accept either the results from lifetime studies in animals or the above battery in assessing the carcinogenicity of biomass products.

There are many nonregulatory applications of short-term tests. Included in these uses are: (1) screening new substances; (2) testing commercial products; (3) responding to occupational health concerns of commercial products; (4) making structure activity relationships with known mutagens/carcinogens; (5) conducting basic research in the biological activity of chemical substances, and (6) risk assessment. For the remainder of this discussion, I will address the first two applications.

Reliable economic and sensitive short-term tests are extremely useful in screening the large number of new compounds synthesized by Hoechst in different areas of chemical research. As no single short-term assay is available which can detect the different types of genotoxic events, several different assays are usually employed in a testing program. Hoechst has a variety of assays available in its own laboratories in Germany or through contract in Europe and in the US.

*Gene mutations* are tested using the Ames assay with and without metabolic activation, reverse mutations in *E. coli* WP2, point mutations in *Saccaromyces cerevisiae,* and Chinese hamster HGPRT forward mutation. *Chromosomal aberrations* are evaluated using in vitro cytogenetics, in vivo cytogenetics, micronucleus, and dominant lethal effects in rodents.

*Primary DNA damage* is assessed in one of the UDS tests (WI-38 human lymphocytes, primary rat hepatocyte, or golden hamster embryo fibroblasts), in Sister chromatid exchange, *B. subtilis* 'rec' assay and DNA binding. Tests for cell transformation utilize the BALB 3T3 cells, Syrian hamster embryo cells, and mouse 10T ½ cell lines.

For screening new compounds, Hoechst often uses a flexible battery consisting of a core of three tests. For example, these may include the Ames assay with and without metabolic activation, Sister chromatid exchange in CHO cells and micronucleus in mice. The Ames test is chosen first since it is the most validated and standardized and has a high degree of correlation with many classes of chemicals that are carcinogens or noncarcinogens. Although the micronucleus test has limitations (it is not a sharp indicator and it does not distinguish clastogenic effects from true aberrations), it nevertheless is easy to perform, rapid, inexpensive and provides an in vivo system to the basic screening program. Sister chromatid exchange is selected since it is reliable and sensitive to many known carcinogens.

Alternate tests are utilized when the core of tests is not appropriate. For example, antibiotics are generally not tested in the Ames assay since many are too toxic to the *Salmonella* organism. In addition, other assays are chosen to answer special questions regarding the compound, such as metabolism, effects on repair processes, induction of cell transformation, DNA-binding potential, and differential effects on somatic cells and germ cells.

It is imperative to maintain flexibility in a battery of tests for the following reasons:

– The properties of a chemical can influence the choice of assay (antibiotics, spindle poisons, or enzyme inhibitors).

– The results of limited existing data can redirect the strategy of further testing.

– Often testing is conducted for internal information only; the number of tests selected is restricted and usually limited to screening.

– Pharmacokinetic differences, when known, may influence the test selection.

– International requirements for new compound development vary.

– When equivocal results are generated on a compound, a more complex battery may be necessary to resolve the differences.

– New information on existing tests and new tests can be easily incorporated into the testing program.

Assays used to test commercial products should be selected on an individual basis depending upon the compound's structure, its metabolism and conditions of use.

The final part of this presentation concerns problems encounterd in the application of short-term tests in safety assessment programs. Decisions based on results from test batteries of each compound should be made from three different levels. First, each assay should be individually evaluated to include the appropriateness of the assay, the use of standardized protocols, and the criteria of evaluating responses, protocol adherence, and the scientific merits of the laboratory and its professional staff as suggested by *De Serres* [1]. Another level of decision involves the overall evaluation of multiple test results. It is more difficult to establish criteria for decision making from multiple tests, since equivocal and mixed responses often cloud the overall interpretation. Various methods of evaluating these data have been proposed, such as the 'weighted contribution method' of scoring as described by *Brusick* [2], which evaluates both negative and positive responses and attempts to consider potency.

The third level involves extrapolation of the biological end point of short-term tests to human health. The relevance of the data from short-term assays to humans is modulated by factors such as differences in DNA repair processes, pharmacokinetics, and physical/chemical barriers to target tissues, molecular dosimetry, and the extrapolation of the results to very low exposure levels in humans. The candidacy of new products in development depends upon continual appraisal of effects in these as well as other toxicological investigations. In general, when analyzing data on a compound, the absence of negative responses in any of the tests should not mislead one to conclude the absence of a carcinogenic risk for man. Likewise, positive responses, in the absence of animal data, should not be used as proof that a compound is a carcinogen for man.

One method for employing these tests in carcinogenicity assessment is the 'decision point approach' as proposed by *Weisburger and Williams* [3]. The approach is composed of four stages of testing interspersed by three decision-making points. The first decision is made after conducting short-term tests and considering the structure of the test substance. The next decision is made after limited bioassays are conducted in vivo. Finally, an evaluation is made based on the preceding data as well as the results of long-term bioassays in animals. Although the applications of the approach have potential, they are limited and cannot always be followed. For example, some compounds developed during the 1970s, without benefit of short-term tests, were found to be oncogenic in animals. Once oncogenic activity is detected, short-term mutagenicity studies can reveal whether this is related to genotoxic or epigenetic mechanisms. This information is valuable because considerations of the mode of activity will facilitate the extrapolation of data from animals to man and from the laboratory to the real world.

A clearer understanding of the meaning and relevance of these short-term tests will become necessary as they play an ever-increasing role in the study of product safety. An assessment of their utility for the safety evaluation of chemicals must consider the predictability of the effects in animals and man and their reliability to give reproducible results. As more sophisticated test systems are utilized, test selection criteria should favor relevance over sensitivity. Negative responses in higher phylogenetic test systems should remove the suspicion of hazard to man, as positive responses give cause for concern. The importance of this is obvious as the results are used to help identify and rank potential carcinogenic substances. One system recently proposed by *Squire* [4] for ranking carcinogens utilizes both positive and negative short-term data as well as other factors. This proposed ranking system provides a score for each of the various factors used in predicting or identifying carcinogens. These factors include the following: number of different species affected, number of histogenetically different types of neoplasms in one or more species, spontaneous incidences in appropriate control groups of neoplasms induced in treated groups, dose-response relationships, malignancy of induced neoplasms and genotoxicity measured in an appropriate battery of tests. The author proposes ranking carcinogens into five classes or categories based on the total of all factor scores. Based on the rank, more sophisticated regulations can decide how these chemicals can or should be used. As short-term tests continue to be used in safety assessment programs, a clearer and more uniform understanding must exist in both government and industry on which tests will most accurately predict the end point desired and the relationship of that end point to human health. Industry and government must continually interact so that the application of existing tests and the development of new ones are more uniform within and between countries. Consistent regulations may result from more rational, logical and uniform testing programs.

In summary, short-term mutagenicity tests are highly useful tools as determinants of specific end points in response to chemical exposure of in vitro and in vivo biological systems. The application of these tests is directed by both regulatory and scientific needs in development and continuing safety assessment programs of commercial products. When the limitations of these tests are viewed in their proper perspective, they can provide useful information at all levels of safety assessment. When the results are evaluated along with appropriate animal and human data, they can provide insight into the possible mechanisms of mutagenicity and carcinogenicity, and in the future provide more certainty than was previously available.

## *References*

1 De Serres, F.J.: Problems associated with the application of short-term tests for mutagenicity in mass-screening programs. Environ. Mutagen. *1:* 203–208 (1979).

2 Brusick, D.: Principles of genetic toxicology, pp. 96–105 (Plenum Press, New York 1980).

3 Weisburger, J.H.; Williams, G.M.: Carcinogen testing: current problems and new approaches. Science *214:* 401–407 (1981).

4 Squire, R.A.: Ranking animal carcinogens: a proposed regulatory approach. Science *214:* 877–880 (1981).

L.S. Grosso, PhD, Merck and Co., P.O. Box 2000, Rahway, NJ 07065 (USA)

# Proffered Papers

Safety Evaluation and Regulation of Chemicals. 1st Int. Conf., Boston, Mass., 1982, pp. 166–172 (Karger, Basel 1983)

## On-Line Data Collection in the Safety Evaluation of Chemicals

*Finis L. Cavender*

The Mitre Corporation, McLean, Va., USA

The impetus to consider the integral use of computers in toxicological studies arose with the promulgation of the Good Laboratory Practice (GLPs) by the Food and Drug Administration. The GLPs were instituted as an attempt to assure that safety evaluation studies are conducted with as uniform and precise procedures as possible. For many laboratories, new levels of instrumentation calibration and overall quality assurance were initiated. A standard operating procedure (SOP) was required for each technique or method utilized in routine studies. Such SOPs allowed a quality assurance audit to verify the use of the correct methodology with properly calibrated instrumentation. GLPs do not forbid one to make a mistake instead, they emphasize the documentation of the exact action taken and thereby allow proper interpretation of the results that follow such action. This emphasis on documentation of protocol, SOPs, and study details has in some ways shifted the emphasis from the 'science' of the study to the 'bookkeeping' of a myriad of details. A need to free the technician to think science has led to a rapid incorporation of computer science in safety evaluation studies [1].

### *Handwritten Data*

Over a sufficiently long period of time, handwritten data will contain approximately 0.3% errors. Coupled with this is the problem of illegibility. Furthermore, these sources of error and confusion are magnified at each step of transcription, verification, statistical or other analyses, and reporting. While this is a relatively low error rate, the final report of a lifetime study in two species of laboratory animals may contain enough errors to question the

*Table I.* Recording positive findings

| Time | Observation | |
|---|---|---|
| 09.00 | all males normal; females 0123, 0125 and 0127 normal; 0124 and 0126 salivation | ABC |
| 09.30 | all males normal; females 0124 thru 0127 salivation; lacrimation, and dyspnea, 0123 normal | ABC |
| 10.00 | all males normal; females 0123 thru 0127 gasping and ataxia | DEF |
| 10.30 | all males salivation; all females salivation, gasping and ataxia | ABC |

When recording positive findings, it is not clear if animal 0124 exhibited salivation at 10.00 h.

interpretation of the stated results. Moreover, an inordinate amount of time is required to check the data and any associated reports to guarantee total quality assurance. Such problems may be circumvented if each data entry is electronically recorded, automatically transferred to a data file in a computer, and utilized directly from the file in the final report. No handwritten entry or hand transcription is required to utilize such data points.

### *Automated Data Collection*

While computers may eliminate the errors associated with the manual collection of data, several potential problems exist. Data security, loss of data, data storage, and computer 'down-time' are the major considerations. These problems can be minimized but not without costly back-up equipment and considerable programming. Other problems include the 'response' time for data entry and the programming necessary to retrieve the data in a useable format.

#### *Format of Toxicology Data*

Prior to GLPs, some laboratories collected data by recording only 'positive' findings as simulated in table I. In such data it is not unequivocal if female 0124 exhibited salivation at the 10.00 h observation or if the technician forgot to look for or record salivation for animal 0124. GLPs require data for each individual animal. In addition, all observations are to be recorded (table II). It is now clear that salivation was not present at the 10.00 h observation for animal 0124.

*Table II.* Individual animal data (animal no. 0124, female)

| Observation | Time | | | | | |
|---|---|---|---|---|---|---|
| | 08.00 | 08.30 | 09.00 | 09.30 | 10.00 | 10.30 |
| No reactions | × | × | NA | NA | NA | NA |
| Salivation | | | × | × | NE | × |
| Lacrimation | | | | × | NE | |
| Dyspnea | | | | × | NE | |
| Gasping | | | | | × | × |
| Ataxia | | | | | × | × |

When all individual animal data are recorded, it is now clear that salivation was not observed at 10.00 h for animal 0124.

Automated data collection can be in any format desired, including that used in manually recording data.

*Automated Collection Systems*

There are three basic types of automated collection systems: (1) instrumentation with printers; (2) computer-assisted 'batch' systems, and (3) computer-assisted 'on-line' systems. Instrumentation systems with printers are common in analytical and clinical chemistry laboratories, e.g., scintillation counters or hematology equipment. Such collectors accurately measure and record sample values and associated parameters, but still require transcription for summarization, statistical analyses, and reporting. Batch systems allow the collection with possible correction of data from one or several locations for a given period of time. Such systems are not flexible since they do not allow 'interaction' between technician and computer, or 'prompting' to notify the user that one or more entries have been overlooked or that the animal weight is not in line with previous entries. However, the collected batch of data can be conveniently transferred to a mainframe computer at a later time by several means including computer tape or floppy discs. On-line data collection requires a direct connection with the mainframe computer or a portable or remote minicomputer. Being on-line means the user and computer can 'talk' to each other, permitting the computer to guide the collection of data. In so doing, there is always a finite 'response' time required for the computer to process all of the questions or entries. This response time is determined by the size of the computer or, more specifically, the 'memory' capacity of the computer. The other serious problem that can arise with on-line collection is computer 'down' time, or rather time during which the computer is not functioning properly.

*Batch versus On-Line Collection*

The decision to choose one system over another depends upon the logistics within a laboratory. Some important considerations are: (1) size of the facility; (2) ability to write and develop computer programs in-house; (3) desired flexibility; (4) the size of budget or cost of the mainframe computer and other related equipment and personnel; and (5) possible expansion or add-on capability at some future date.

*Size of Facility*

The number of direct trunk lines connecting to the mainframe computer usually determines which approach to take. With any given computer, a greater number of trunk lines increases the response time for any one user. This can be offset by a large memory capacity which allows the computer to work faster. The other aspect of this question deals with the need to interact or render judgement for a given observation. If the technician does not have to record an observation taken from a list of several possible choices then a batch system would be appropriate: simpler and less costly. However, if the technician must decide if the observation is a discharge from the eyes, nose, or mouth of a small rodent, programming for such would be more costly. In addition, interaction with the computer and/or the entry of additional commentary may be appropriate. If, for any one type of observation, an appropriate batch collection system is available, one might chose the batch system for an entire facility as shown in figure 1a. An on-line system with a limited number of trunk lines is shown in figure 1b. For a large facility, with portable or remote on-line minicomputers, the technician is on-line with the remote unit which acts as a batch collection system, since the data are transferred to the mainframe computer by floppy disc, etc. Such a system is illustrated in figure 1c. An additional modification presently in use is the recording of data 'batch-wise' on a tape recorder which is later entered on-line by a technician. This type of system may be justified when one does not want a pathologist or other key scientist to be delayed by the response time of the computer or when the cost of the equipment necessary to be on-line is not justified. This system is illustrated in figure 1d.

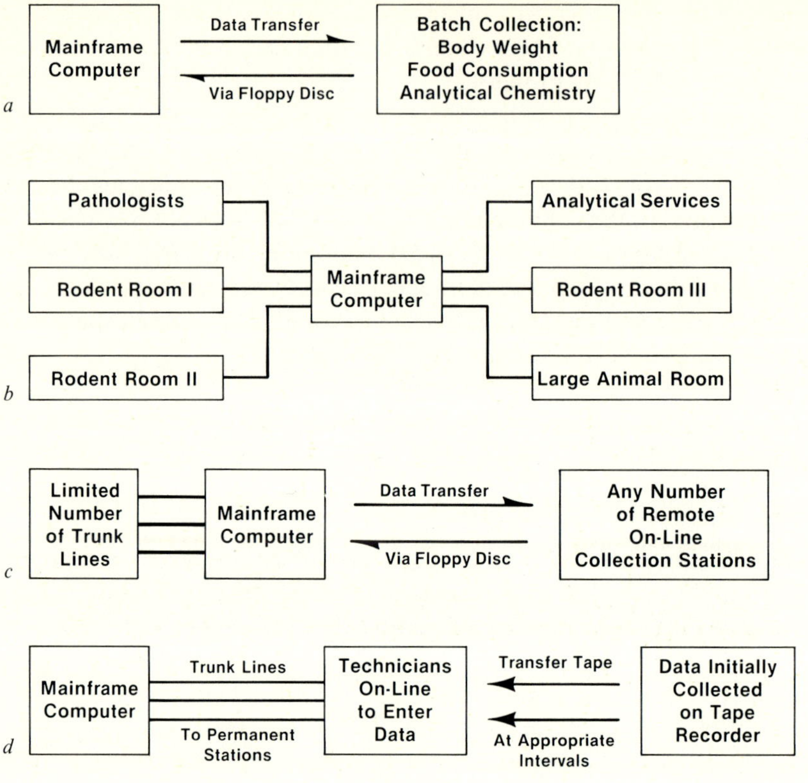


*Fig. 1.* *a* Laboratory data collected using batch collectors. *b* Laboratory data collected using direct connections to the mainframe computer. *c* Laboratory data collected on-line using remote minicomputers. *d* Laboratory data collected on recording devices and later transferred on-line to mainframe computer.

### *Entering Data*

Data are usually entered into batch or on-line systems by typewriter keyboard, touch of the appropriate entry displayed on a CRT screen, or via a barcode reader. Barcode readers are error-free entry systems but they require the ability to generate one's own barcode labels. Keyboard or touch entries usually depend on the type of programming chosen. Keyboard entry often accompanies a question/answer format, i.e., the program proceeds through a sequence of questions which require a response from the user before proceeding to the next question. This may be simple, but is often time consuming. Touch systems require more complex programming but usually follow a 'form' format. The form can be identical to the printed forms previously used in manual data collection. A touch system can be modified to allow the user to type in an observation in the appropriate space on the screen.

### *Designing the System*

Safety Evaluation Studies can usually be divided into at least 3 phases. Phase I includes the identifying information for the study and constitutes the initialization of the study. Phase I would identify: (1) the study director; (2) the test system; (3) the test article; (4) the preliminary schedule; (5) the technicians involved; and (6) any additional information appropriate to the study. Phase II includes the details of the protocol and would clearly identify: (1) test groups; (2) animal numbers; (3) in-life observations; (4) sacrifice-related parameters; (5) necropsy procedures; (6) pathology procedures; and (7) statistical and other processing parameters. These phases will be completed and verified prior to the start of the study and would include such preparatory steps as printing barcode labels, etc. Phase III would include the collection and processing of the study data from the quarantine period to the final report.

### *What to Do First?*

Considering the size of your facility and the flexibility required in processing and reporting study results, one must first identify a person or persons who can rationally decide what type of system would work best for you. If possible, visit and observe in operation some of the different types of systems currently in use.

While there are many systems, these working facilities may be a place to start:

– A Direct On-Line System (fig. 1b) – Wil Research Laboratories, Inc., Cincinnati, Ohio.

– A Remote On-Line System (fig. 1c) – ToxiGenics Inc., Decatur, Il.

– A Modified On-Line System (fig. 1d) – Experimental Pathology Laboratories Inc., Herndon, Va.

The apparent costs are high and may be very important in determining your direction; however, consider the following [2]:

(1) The actual technician time saved via on-line data collection over the manual collection of data is approximately 25% in man-hours.

(2) The manpower requirements for the quality assurance are reduced to occasional 'system checks' of the programs (note: for chronic studies, this may represent a saving of one man-year).

(3) The tabulated results for the final report for studies including teratology studies and studies with considerable pathology are available immediately, rather than 6 or more months following study termination.

## *References*


1 Phifer, D.C.: Why technician training is important. Lab. Animal *9:* 43–45 (1980).

2 Cavender, F.L.: Churukin, P.V.; Gralla, E.J.: A comparison of on-line data collection system and manual collection of data for inhalation toxicology studies. Toxicologist *1:* 152 (1981).


Finis L. Cavender, PhD, The Mitre Corporation, Hazardous Substances Evaluation, 1820 Dolly Madison Boulevard, McLean, VA 22102 (USA)

Safety Evaluation and Regulation of Chemicals. 1st Int. Conf., Boston, Mass., 1982, pp. 173–182 (Karger, Basel 1983)

# Carcinogen Specificity of Gamma-Glutamyl Transpeptidase-Rich Foci Chemically Induced in Hamster Buccal Pouch Epithelium[1]

*D.P. Bernard, D.B. Solt*[2]

Harvard School of Dental Medicine, Departments of Periodontology and Oral Pathology, Boston, Mass., USA

The development of a simple technique for the early detection and enumeration of precancerous cells in squamous epithelium would be of great theoretical and practical significance. Hamster buccal pouch epithelium (HBPE) may serve as an ideal experimental tissue for the development of such a technique. This carcinogen-susceptible epithelium is of uniform thickness and lacks hair follicles or other complex adnexal structures. After a brief nondestructive exposure to a chemical carcinogen, the entire pouch epithelium can be detached from the underlying connective tissue and mounted intact on glass slides for histochemical staining and microscopic examination of the basal layer. Using this approach, and the histochemical marker gamma-glutamyl transpeptidase (GGT), we have detected discrete nests of carcinogen-altered basal cells which may be precancerous in nature. GGT has been detected in a variety of human cancers [7, 8, 10] and has also served as a useful histochemical marker for carcinogen-induced hepatomas and presumptive precancerous lesions in rodent liver [1, 6, 12, 15, 17, 20, 32].

GGT-rich foci form in HBPE during a brief series of topical applications of the potent carcinogen 7,12-dimethylbenz(a)anthracene (DMBA). These foci are rapidly induced, ostensibly clonal in nature, and persistent [33]. These findings and the observation that a significant proportion of DMBA-induced pouch carcinomas also contain GGT-stained cells [31], support the hypothesis that the carcinogen-induced, GGT-rich foci are precancerous in nature.

[1] Supported by Grant CA 28620 from the National Cancer Institute.

[2] Recipient of Research Career Development Award CA 00673 from NIH.

In this study we screened several noncarcinogenic irritants and five additional chemicals known to induce cancer in rodent epithelium, for their ability to induce GGT-rich foci in HBPE. The irritants were selected because of their known ability to cause hyperplasia and inflammation in rodent epithelium. The results add further support to our hypothesis that the chemical induction of carcinomas in HBPE is preceded by the early induction of GGT-rich foci which serve as material precursors for subsequent tumor development.

## *Materials and Methods*

### *Animals and Treatment*

Noninbred male Syrian golden hamsters, 4–6 weeks of age, were used in all but one of the experiments. These animals were obtained from Charles River Breeding Laboratories, Inc., Wilmington, Mass. For one experiment in which 20-methylcholanthrene was applied for up to 290 days, both male and female BIO F1D Alexander hamsters were used. These hybrid hamsters were generously donated by Bio-Research, Inc., Cambridge, Mass. All animals were housed 4–6 per plastic cage with wood chips (Pine Dri, St. Regis Paper Co., Garfield, N.J.) for bedding. The animals were maintained in a controlled environment with an alternating 12-hour light-dark cycle with water and laboratory chow (Rat Chow 5012; Ralston-Purina Co., St. Louis, Mo.) ad libitum. At the conclusion of each treatment period the animals were sacrificed with carbon dioxide and buccal pouch epithelium was removed to prepare whole mounts suitable for GGT histochemical staining.

### *Experimental Regimens Involving Application of Carcinogens*

(a) 7,12-Dimethylbenz(a)anthracene (DMBA, Sigma Chemical Co.). The experimental technique of DMBA treatment has been reported previously [31]. A 0.5% solution of DMBA in mineral oil was applied by painting the entire mucosal surface of both pouches each Monday and Friday for 4 consecutive weeks. The animals were sacrificed 1 week after the last application. Control animals were similarly treated with mineral oil.

(b) Methylbenzylnitrosamine (MBN, Ash Stevens Inc.). 18 animals received a 5% solution of MBN in propylene glycol, applied by painting both pouches each Monday, Wednesday and Friday for 4 weeks, and then each Monday and Friday for 3 additional weeks; 4 animals were sacrificed after the first 4 weeks of treatment and pouch epithelial whole mounts were prepared for GGT histochemistry. The remaining animals were maintained without further treatment and sacrificed when moribund or in the 22nd week for histological evaluation and enumeration of epithelial pouch tumors. Control animals were similarly treated with propylene glycol.

(c) 20-Methylcholanthrene (MC, Sigma Chemical Co.). For experiment 1, 0.1 ml of a 0.5% solution of MC in acetone was deposited in the depth of both buccal pouches each Monday, Wednesday and Friday for 9 weeks. In experiments 2 and 3, a 0.5% solution of MC in mineral oil was applied by painting both buccal pouches each Monday, Wednesday and Friday for 41 and 16 weeks, respectively. Control animals were treated in an equivalent manner with acetone or mineral oil.

(d) Methylnitrosourea (MNU, ICN Pharmaceuticals). A 1% solution of MNU was prepared in citrate buffer (pH 6.0) containing 25% ethanol; 0.5 ml of this solution was administered by a single injection into the loose submucosal connective tissue superficial to each buccal pouch. Control animals were similarly treated with the citrate buffer-ethanol solution alone. These animals were sacrificed 4 days after the submucosal injection.

(e) Benzo(a)pyrene (BP, Sigma Chemical Co.). A 0.5% solution of BP in mineral oil was applied by painting the pouch epithelium each Monday, Wednesday and Friday for 15 consecutive weeks. Control animals were similarly treated with mineral oil.

*Experimental Regimens Involving Application of Noncarcinogenic Irritants*

(a) Turpentine (Sterling-Clark-Lurton Corp., Malden, Mass.). A 50% solution of turpentine in mineral oil was administered by painting both buccal pouches daily for 30 consecutive days.

(b) Acetic acid. 0.1 ml of a 25% aqueous solution of acetic acid was delivered to the depth of both buccal pouches each Monday and Friday for 6 weeks. Control animals received an equivalent volume of water.

(c) Ethylphenylpropiolate (EPP, Aldrich Chemical Co.). 0.1 ml of a 6.6% solution of EPP in acetone was deposited into both buccal pouches each Monday and Friday for 6 weeks. Control animals were similarly treated with acetone.

(d) Croton oil (Sigma Chemical Co.). Undiluted croton oil was painted on the entire mucosal surface of both buccal pouches daily for 30 consecutive days.

(e) Phenol (Fisher Chemical Co.). A 5% solution of phenol in propylene glycol was applied by painting both buccal pouches each Monday, Wednesday and Friday for 4 weeks. Control animals were painted thrice weekly with propylene glycol alone.

(f) 12-0-tetradecanoylphorbol-13-acetate (TPA, Chemical Carcinogenesis, Inc., Eden Prairie, Minn.). 0.1 ml of a 0.1% solution of TPA in acetone was applied to the depth of both buccal pouches each Monday, Wednesday and Friday for 4 weeks. Control animals received an equivalent volume of acetone in like fashion.

(g) Vitamin A palmitate (Sigma Chemical Co.). A 10% solution of vitamin A palmitate in paraffin oil was painted on the mucosal surfaces of both buccal pouches every Monday, Wednesday and Friday for 7 consecutive weeks.

*Whole Mount Preparation and GGT Histochemistry.* At the completion of each experiment, the animals were sacrificed with a lethal dose of carbon dioxide and whole mounts of pouch epithelium were prepared as described previously [31, 33]. The pouches were excised, opened and placed in an ice-cold aqueous solution of 1% acetic acid with constant stirring for 1 h. Large pieces of the epithelium were detached from the underlying connective tissue and mounted basal layer up on microscopic slides. After air drying, the full thickness whole mounts were secured to the glass surface with glycerin gelatin. Localization of GGT activity was accomplished as previously described [33] using the histochemical method of *Rutenburg* et al. [23].

*Quantitation of GGT-Stained Foci.* In order to quantitate GGT-stained lesions, the outline of each whole mount was traced onto graph paper with square millimeter divisions. The number of GGT-stained foci was determined microscopically and expressed in terms of surface area.

*Histology.* Selected pouches were fixed in formalin and processed for routine histopathological evaluation with HE staining. Cheek pouch, forestomach and esophageal tissues from MBN-treated animals were treated in a like manner.

*Table I.* Quantitation of GGT-stained foci induced in HBPE by carcinogens

| Agent | Number of experimental animals / Number of control animals | Maximum number of applications[a] / Experiment duration, days | Number of plaques/ 10 $cm^2$ |
|---|---|---|---|
| MBN | 4/2 | 18/28 | 280 ± 162[b] (4 of 4)[c] |
| DMBA | 4/2 | 8/35 | 92 ± 54 (4 of 4) |
| MNU | 4/2 | 1/4 | 43 ± 47 (3 of 4) |
| BP | 3/1 | 45/116 | 3 ± 2 (3 of 3) |
| MC[d] | | | |
| Experiment 1 | 7/3 | 26/66 | 0.2 ± 4 (1 of 7) |
| Experiment 2 | 16/8[e] | 90/290[f] | 0.1 ± 0.3 (4 of 16) |
| Experiment 3 | 3/1 | 45/116 | 0.2 ± 0.3 (1 of 3) |

[a] See ‘Materials and Methods’ for details of treatment.
[b] Mean ± standard deviation. Values shown for carcinogen-treated animals. Values for vehicle-treated animals were 0 in all cases.
[c] Proportions of experimental animals with GGT-stained foci shown in parentheses.
[d] Experiment 1 employed 0.5% MC in acetone. Experiments 2 and 3 employed 0.5% MC in mineral oil.
[e] BIO F1D Alexander hamsters were used in this experiment only.
[f] Groups of 2, 2, 4 and 8 animals were sacrificed on days 37, 140, 210 and 290, respectively.

## *Results*

### *Carcinogen Induction of GGT-Stained Foci*

3 of the 5 carcinogens studied induced numerous foci of GGT positivity in the treated pouch epithelium (table I). The DMBA treatment regimen resulted in the production of discrete nests of cells which stained intensely with the GGT reaction product (fig. 1). These GGT-stained cells were indistinguishable from those previously described [33]; 1 week after completion of the 4-week regimen the largest DMBA-induced foci were plaques ranging up to approximately 0.5 mm in maximum dimension; 4 weeks of MBN treatment resulted in the production of GGT-stained foci which were similar in size, number and staining intensity to the DMBA-induced lesions. The single submucosal induction of MNU also resulted in the production of GGT-positive foci. The majority of the MNU-induced foci consisted of individual cells and cell doublets. The largest lesion in this group consisted of a nest of approximately 8 cells.

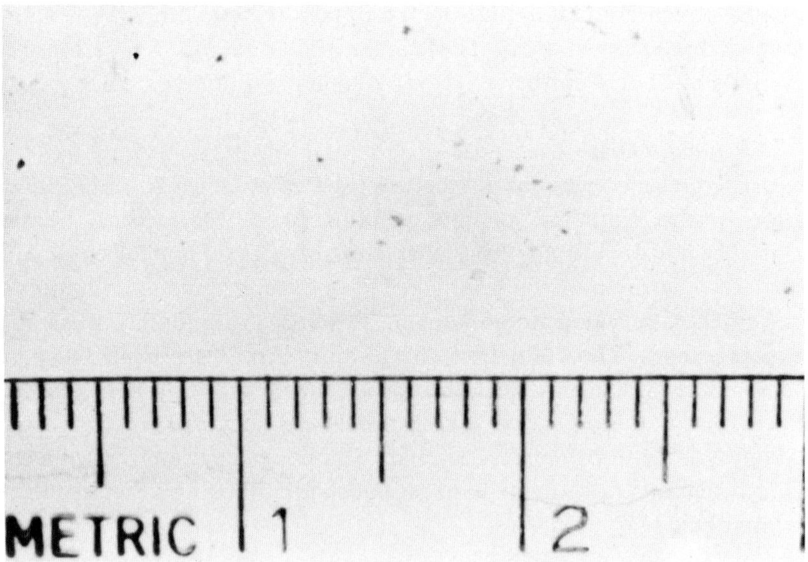


*Fig. 1.* Portion of whole mount of DMBA-treated HBPE stained for GGT histochemical activity. Multiple dark foci represent discrete populations of GGT-stained cells resulting from 4 weeks of carcinogen treatment.

Application of BP for a period of 116 days resulted in the production of very few GGT-positive foci (table I). No tumors formed in any of the 3 BP-treated animals.

MC was applied to the pouch epithelium in three experiments ranging in duration from 66 to 290 days. In each case the yield of GGT-stained foci was quite low. In experiment 2 hybrid hamsters known to be susceptible to MC-induced mammary cancers [14] were employed. The pouch epithelium of these animals was no more susceptible to the induction of GGT-staining foci than the pouch epithelium of the noninbred animals used in experiments 1 and 3. No pouch tumors developed in any of the MC-treated animals.

*Lack of GGT Staining in Pouch Epithelium Treated with Noncarcinogenic Irritants*

With the possible exception of TPA, the noncarcinogenic irritants examined in this study failed to produce histochemical GGT activity in the treated pouch epithelium. Of the 4 animals treated with TPA, 2 showed a total of three stained areas. These areas, unlike the discrete lesions seen in carcinogen-treated epithelia, were ill-defined and faintly stained.

*Histopathologic Alterations*

The present study did not include a systematic histopathological evaluation of the epithelial changes resulting from the various treatments. However, a histopathological examination of selected pouches from animals treated with ethylphenylpropiolate, TPA, croton oil, acetic acid, turpentine, and phenol, revealed no convincing evidence of epithelial thickening or hyperkeratosis. The epithelium of vitamin A-treated pouches displayed a generalized cytoplasmic vacuolization and necrosis of cells in the basal layer. Large, discrete, submucosal aggregates of lymphocytes were also seen in the vitamin A-treated pouches. Similar lymphocytic lesions have been previously demonstrated in the hamster pouch after feeding a diet rich in 13-cis-retinoic acid [11].

*Tumor Production with Methylbenzylnitrosamine*

Because of the high efficiency with which MBN induced GGT-rich foci in the hamster pouch epithelium, we examined the carcinogenic influence of MBN on this tissue. The carcinogenic regimen employed was highly toxic, and the experiment was terminated after 22 weeks as only 5 of the original 18 animals were still alive. Each of the 5 animals had between 2 and 10 epithelial cheek pouch tumors including one or more well-differentiated squamous cell carcinomas. All animals sacrificed after the seventh week of this experiment also contained multiple squamous papillomas of the forestomach. One squamous cell carcinoma of the forestomach was also encountered in this group.

## *Discussion*

The results of this study suggest that the chemical induction of GGT-rich foci in HBPE is an early manifestation of carcinogenesis rather than one of nonspecific irritation. Both DMBA and MBN efficiently produced GGT-rich lesions during the first 4 weeks of treatment. It is well known that DMBA is a potent carcinogen for this tissue [24, 25]. This study demonstrates that MBN is also a potent carcinogen for HBPE. All MBN-treated animals surviving for 17 weeks had squamous cell carcinomas of the buccal pouch.

The carcinogenicity of the direct-acting carcinogen, MNU, has been reported for a broad range of tissues and species including the buccal pouch of the Syrian hamster [4]. In this study, MNU was administered by the submucosal route and was found to be highly efficient at producing GGT-positive foci within 4 days after a single injection. In an unpublished preliminary study, we were not successful in producing GGT-positive foci with topical application. This disparity may have resulted from a rapid inactivation or inadequate penetration of the carcinogen.

Long-term exposure to both BP and MC induced relatively few GGT-positive lesions and no neoplasia. In a previous report by *Shklar* et al. [26], the buccal pouches of 30 Syrian hamsters were treated thrice weekly with BP for 24 weeks with no gross or microscopic evidence of cancer formation. The lack of susceptibility of HBPE to both BP and MC are surprising in view of the carcinogenic potency of these agents in mouse skin. Perhaps the most likely explanation for this disparity is a relative inability of pouch tissue to enzymatically convert these agents to an ultimate carcinogenic metabolite.

We tested 7 noncarcinogenic irritants in an attempt to determine whether the induction of GGT-positive foci in HBPE is a consequence of irritation unrelated to the process of carcinogenesis. These agents were selected because of their known ability to induce inflammatory hyperplasia in mouse skin [5, 9, 13, 19, 21, 22, 30], hamster skin and/or HBPE [2, 3, 16, 18, 27, 28]. However, at the completion of each experiment a histopathological examination of pouch specimens from the animals treated with noncarcinogens revealed no obvious hyperplasia. This surprising lack of hyperplasia may have resulted from the acquisition of tissue resistance to the toxic agents. Indeed, *Sisskin and Barrett* [28] have shown that hamster epidermis rapidly overcomes the hyperplastic effect of TPA and perhaps a similar phenomenon applies to HBPE exposed to TPA and the other irritants used in this study. With the possible exception of TPA the irritants examined in this study did not give rise to GGT-positive foci. TPA produced a total of only 3 GGT-positive areas in 4 animals.

There have been numerous reports of enhanced GGT activity in human tumors of a variety of tissues including esophagus [8], colon [7], urinary bladder and prostate [8] and liver [10]. There is also compelling evidence that chemically induced liver cancers, most of which are GGT staining, evolve from GGT-rich precancerous lesions [1, 17, 20, 32]. The GGT-rich foci described in this study may also represent cell populations undergoing malignant evolution. As previously shown [33], the GGT-rich foci induced in HBPE using DMBA are rapidly induced, focal and apparently clonal in origin, and persistent. Moreover, both agents known to be carcinogenic for HBPE were by far the most efficient inducers of GGT-rich lesions. Additionally, the HBPE cancers induced by these agents occasionally contain GGT-rich cells.

In principle, the employment of epithelial whole mounts for the early detection and quantification of precancerous cell populations should be a rapid and biologically relevant means of identifying carcinogenic agents. Perhaps the sensitivity of this approach could be increased by the development of additional carcinogen-susceptible hamster strains. Also, pretreatment with inducers of the appropriate activating enzymes [29] may broaden the spectrum of carcinogenic agents which produce GGT-rich precancerous foci in HBPE.

## *References*


1 Cameron, R.; Kellen, J.; Kolin, A.; Malkin, A.; Farber, E.: Gamma-glutamyl transferase in putative premalignant liver cell populations during hepatocarcinogenesis. Cancer Res. *138:* 823–829 (1978).

2 Cavalaris, C.J.; Krikos, G.A.: Vitamin A-produced mucous metaplasia. J. oral Ther. Pharmacol. *3:* 452–461 (1967).

3 Craig, G.T.; Franklin, C.D.: The effect of turpentine on hamster cheek pouch mucosa. A model of epithelial hyperplasia and hyperkeratosis. J. oral Path. *6:* 268–277 (1977).

4 Edwards, M.B.; Germain, J.P.: Oral carcinogenesis by intraperitoneal N-methyl-N-nitrosourea in the syrian hamster. Archs oral Biol. *2:* 631–633 (1980).

5 Evensen, A.: Effects of carcinogens on cell proliferation. Acta path. microbiol. scand., suppl. 156, pp. 93–156 (1962).

6 Fiala, S.; Mohindru, A.; Kettering, W.G.; Fiala, A.E.; Morris, H.P.: Glutathione and gamma-glutamyl transpeptidase in rat liver during chemical carcinogenesis. J. natn. Cancer Inst. *57:* 591–598 (1976).

7 Fiala, S.; Trout, E.; Pragani, B.; Fiala, E.S.: Increased gamma-glutamyl transferase activity in human colon cancer. Lancet *i:* 1145 (1979).

8 Fiala, S.; Trout, E.C., Jr.; Teague, C.A.; Fiala, E.S.: Gamma-glutamyltransferase, a common marker of human epithelial tumors? Cancer Det. Prev. *3:* 471–485 (1980).

9 Frei, J.V.; Stephens, P.: The correlation of promotion of tumor growth and induction of hyperplasia in epidermal two-stage carcinogenesis. Br. J. Cancer. *22:* 83–92 (1968).

10 Gerber, M.A.; Thung, S.N.: Enzyme patterns in human hepatocellular carcinoma. Am. J. Path. *98:* 395–400 (1980).

11 Gilmore, W.; Giunta, J.L.: The effect of 13-cis-retinoic acid on hamster buccal pouch carcinogenesis. Oral Surg. *51:* 256–265 (1981).

12 Harada, M.; Okabe, K.; Shibata, K.; Masuda, H.; Miyata, K.; Enomoto, K.: Histochemical demonstration of increased activity of gamma-glutamyl transpeptidase in rat liver during hepatocarcinogenesis. Acta histochem. cytochem. *9:* 168–179 (1976).

13 Hennings, H.; Boutwell, R.K.: Studies on the mechanism of skin tumor promotion. Cancer Res. *30:* 312–320 (1970).

14 Homburger, R.; Adams, R.A.; Soto, E.; Van Dongen, C.G.: Susceptibility and resistance to chemical carcinogens in inbred syrian hamsters. Prog. exp. Tumor Res., vol. 24, pp. 215–221 (Karger, Basel 1979).

15 Kelengayi, M.M.; Ronchi, G.; Desmet, V.J.: Histochemistry of gamma-glutamyl transpeptidase in rat liver during aflatoxin $B_1$-induced carcinogenesis. J. natn. Cancer Inst. *55:* 579–588 (1975).

16 Lawrence, D.J.; Bern, H.A.; Steadman, M.G.: Vitamin A and keratinization studies in the hamster cheek pouch. Ann. Otol. Rhinol. Lar. *69:* 645–660 (1960).

17 Pitot, H.C.; Barsness, L.; Goldsworthy, T.; Kitagawa, T.: Biochemical characterization of stages of hepatocarcinogenesis. Nature, Lond. *271:* 456–458 (1978).

18 Polliack, A.; Levij. I.S.: Increased incidence of carcinoma induced by DMBA in the hamster cheek pouch in response to vitamin A. Nature, Lond. *216:* 187–188 (1967).

19 Pound, A.W.: Carcinogenesis and cell proliferation. N. Z. med. J. *67:* 88–98 (1968).

20 Pugh, T.D.; Goldfarb, S.: Quantitative histochemical and autoradiographic studies of hepatocarcinogenesis in rats fed 2-acetylaminoflourine followed by phenobarbital. Cancer Res. *38:* 4450–4457 (1978).

21 Raick, A.N.: Ultrastructural, histological, and biochemical alterations produced by 12-O-tetradecanoyl-phorbol-13-acetate on mouse epidermis and their relevance to skin tumor promotion. Cancer Res. *33:* 269–286 (1973).

22 Raick, A.N.; Burdzy, K.: Ultrastructural and biochemical changes induced in mouse epidermis by a hyperplastic agent, ethylphenylpropiolate. Cancer Res. *33:* 2221–2230 (1973).

23 Rutenburg, A.H.; Kim, H.; Fischbein, J.W.; Hanker, J.S.; Wasserkrug, H.L.; Seligman, A.M.: Histochemical and ultrastructural demonstration of gamma-glutamyl transpeptidase activity. J. Histochem. Cytochem. *17:* 517–526 (1969).

24 Salley, J.J.: Experimental carcinogenesis in the cheek pouch of the syrian hamster. J. dent. Res. *33:* 253–262 (1954).

25 Salley, J.J.: Histologic changes in the hamster cheek pouch during early hydrocarbon carcinogenesis. J. dent. Res. *36:* 48–55 (1957).

26 Shklar, G.; Schwartz, J.; Frim, D.: Comparison of DMBA and BP as hamster buccal pouch carcinogens. J. dent. Res. *60:* 647 (1981).

27 Silberman, S.; Shklar, G.: The effect of a carcinogen (DMBA) applied to the hamster buccal pouch in combination with croton oil. Oral Surg. *16:* 1344–1355 (1963).

28 Sisskin, E.E.; Barrett, J.C.: Hyperplasia of Syrian hamster epidermis induced by single but not multiple treatments with 12-0-tetradecanoylphorbol-13-acetate. Cancer Res. *41:* 346–350 (1981).

29 Slaga, T.J.: Carcinogenesis. A comprehensive survey, vol. 5, pp. 1–262 (Raven Press, New York 1980).

30 Slaga, T.J.: Bowden, T.G.; Boutwell, R.K.: Acetic acid, a potent stimulator of mouse epidermal macromolecular synthesis and hyperplasia but with weak tumor-promoting ability. J. natn. Cancer Inst. *55:* 983–987 (1975).

31 Solt, D.B.: Localization of gamma-glutamyl transpeptidase in hamster buccal pouch epithelium treated with 7,12-dimethylbenz(a)anthracene. J. natn. Cancer Inst. *67:* 193–200 (1981).

32 Solt, D.B.; Medline, A.; Farber, E.: Rapid emergence of carcinogen-induced hyperplastic lesions in a new model for the sequential analysis of liver carcinogenesis. Am. J. Path. *88:* 595–618 (1977).

33 Solt, D.B.; Shklar, G.: Rapid induction of gamma-glutamyl transpeptidase-rich intraepithelial clones in 7,12-dimethylbenz(a)anthracene-treated hamster buccal pouch. Cancer Res. *42:* 286–291 (1982).

D.B. Bernard, DMD, Department of Periodontology, Harvard School of Dental Medicine, 188 Longwood Avenue, Boston, MA 02115 (USA)

Safety Evaluation and Regulation of Chemicals. 1st Int. Conf., Boston, Mass., 1982,
pp. 183–190 (Karger, Basel 1983)

# Acute Ethanol and Stress Alteration of Chemical Toxicity[1]

*D.R. Brown[a], H. Chung[b], M.S. Bogdanffy[a]*

[a] Toxicology Program, Northeastern University, Boston, Mass.; [b] Department of Pharmacology, Walter Reed Institute of Research, Washington, D.C., USA

Regulation of chemical safety is based on the responses of sensitive members of the population at risk. Therefore, factors which alter the response of this population to toxic compounds are important. Acute ingestion of alcohol and acute stress are two factors commonly found in the general population which can be shown to affect sensitivity to drugs, chemical exposures and potential carcinogens [13]. Studies have shown that hormones, especially the steroids, can alter mixed function oxidase enzymes in vivo and in vitro [8]. We have been investigating the potential for interactions between alcohol and stress as factors which alter the actions and metabolism of several xenobiotics. Our studies suggest that there is an inhibition of drug metabolism by both acute alcohol and acute stress and that the factors are additive and possibly mediated through the adrenals.

Several studies have shown that the chronic application of stress or the chronic ingestion of alcohol alters the duration and intensity of action of several drugs [6, 10]. These actions are correlated with the activity of the mixed function oxidase (MFO) enzymes of the liver. Few studies have examined the duration or intensity of drug actions under acute administration of stress or alcohol, although this is a more common phenomenon than the chronic exposures. This report is an overview of a series of studies in our laboratory on the effects of alcohol and stress on drug action. It is proposed that acute alcohol and stress act through a common mechanism which is mediated, in part, by the adrenal gland.

[1] This research has been supported in part by grants from the Pharmaceutical Manufacturers Research Grant and BCMG Grant No. RR07143, Department of Health and Human Services.

*Table I.* Treatment protocol for alcohol-stress studies

| Procedure | | Time, min |
|---|---|---|
| A | Ethanol administration | 0 |
| B | HLL application or corticosterone administration[1] | 1 |
| C | Removal of HLL | 60 |
| D | Sacrifice or hexobarbital treatment | 61 |

[1] 45 min after alcohol in noted studies.

## *Material and Methods*

### *Treatment*

Adult male Sprague-Dawley rats were maintained in a temperature-controlled room with a 12-hour light-dark cycle for a minimum of 4 days before they were used in these studies. The rats were fed rat chow and water ad libitum. All experiments were scheduled so that animals were treated between 8 and 10 a.m.

Ethanol (3 or 5 g/kg) was administered in a 15% solution by oral intubation. Controls received equivalent volumes of water. Stress was applied by unilateral hindleg ligature (HLL) with a rubber band, 7–8 cm in circumference, wrapped three times around a hindleg. The ligature was removed at 1 h or at specified times and animals sacrificed or treated as indicated. In the stress-alcohol study, alcohol was administered immediately prior to HLL. Adrenalectomized rats were maintained on rat chow and a solution of 0.9% NaCl plus 0.1% KCl as drinking water ad libitum prior to use. Rats were used for studies no sooner than day 4 after surgery. Corticosterone (Sigma Chemical Co.) 12.5 mg/kg in propylene glycol, i.p., was administered 60 min before sacrifice or as specified. The treatment protocol is shown in table I.

### *Tissue Preparation*

The animals were sacrificed by decapitation, the livers rapidly excised and placed in 1.15% KCl. The livers were blotted dry, weighed, minced into fine pieces and homogenized in 3 volumes of 1.15% KCl with a motor-driven coaxial Teflon pestle and glass homogenizer for 3 min at 1,000 rpm. The homogenate was centrifuged at 9,000 *g* (average) for 30 min and either used directly, or further centrifuged at 78,000 *g* (average) for 90 min to separate microsomes and cytosol. The microsomal pellets were washed once by resuspension in 1.15% KCl and centrifuged at 78,000 *g* (average) for 60 min. The washed microsomal pellets were resuspended in pH 7.4 KCl:Tris buffer (150 m*M* KCl and 50 m*M* Tris).

### *Metabolism Studies*

Experiments were performed on fresh tissue preparations unless otherwise indicated. When frozen tissue was used, samples of 9,000 *g* fractions were 'shell frozen' using acetone and dry ice and stored at –80 °C until used. Incubation conditions used a standard reaction mixture consisting of 2 μmol NADP, 10 μmol $MgCl_2 \times 6H_20$, 50 μmol nicotinamide and 20 μmol glucose-6-phosphate in 25 ml Erlenmeyer flasks. 1 ml liver fraction = 333 mg of wet tissue, substrate and a volume of buffer was added to the flask to make a final reaction volume of 4 ml. Samples were shaken in a Dubnoff metabolic incubator at 37 °C. Plasma hexobarbital was measured by the method of *Cooper and Brodie* [7]; in vitro metabolism of hexobarbital was measured using tritiated hexobarbital by a modification of the method of *Kupfer and Rosenfeld* described by *Chung and Brown* [4, 5]. Aminopyrine-N-demethylase activity in vitro was measured by the method of *Soyka* [15], and aniline hydroxylase activity in vitro was measured by the method of *McMahon* [11]. Benzo(a)pyrene (BAP) metabolism was measured by a modification of the method of *Kinoshita* et al. [9]. Microsomes were prepared from 3-methylcholanthrene (3-MC) pretreated rats according to *Kinoshita* et al. [9] and BAP disappearance was measured by HPLC on a reverse phase $C_{18}$ column. BAP was eluted isocratically using 100% methanol (flow rate 2.2 ml/min). Detection of BAP was at 285 nm by comparison with sample peaks from standard solutions.

Ethanol was determined by an enzymatic assay using an ethyl alcohol analytical kit (Sigma). The corticosterone in plasma was measured by a fluorometric method according to *Zenker and Berstein* [17].

## *Findings*

### *Hexobarbital Metabolism: Effect of Acute Alcohol and Acute Stress*

Correlation between drug pretreatment effects on the duration of action of hexobarbital, on the liver MFO activity and on the pharmacokinetics of hexobarbital distribution have shown the mode of action of many drug interactions. In order to determine whether acute alcohol and acute stress act through a MFO system, in vitro metabolism of hexobarbital was measured and compared with hexobarbital sleep time and the plasma levels of hexobarbital. Adult male rats were treated according to the standard protocol described in the Methods section. The findings are presented in tables II–IV.

It was considered possible that the treatments released an inhibitor of MFO activity into the cell sap. To test this, cytosol from treated rats was added to microsomes of normal rats. No differences in microsomal hexobarbital hydroxylase activity were found when liver cytosol from stress (1 h HLL), alcohol pretreated (3 g/kg, 1 h before sacrifice) or alcohol- and stress-treated rats was mixed with microsomes from control rats. Thus, the inhibition is not due to the presence of an inhibitor in the cytosol. Microsomal P450 levels were decreased in rats pretreated with acute alcohol, acute stress or the combination of alcohol with stress [2] (data not shown).

The plasma corticosterone levels were inversely correlated with the hexobarbital hydroxylase activity in rats pretreated with alcohol, stress or the combined alcohol and stress. These findings suggest a direct correlation between the plasma level of corticosterone and inhibition of hexobarbital hydroxylase. Adrenalectomy reduced corticosterone levels nearly 10-fold but

*Table II.* Effect of acute stress and acute alcohol pretreatment on hexobarbital tissue levels, duration of action and in vitro liver metabolism in the rat [from ref. 3 and 5]

| Treatment | % inhibition of hexobarbital metabolism | Sleep time min | Hexobarbital level at awakening | | |
|---|---|---|---|---|---|
| | | | brain | plasma | brain/plasma |
| Control | 0 | 35.1 ± 3.9 | 44.1 ± 4.7[1] | 66.0 ± 6.8[1] | 0.67 |
| Ethanol (3 g/kg 15%) | 45[2] | 55.1 ± 13.6[2] | 26.9 ± 2.6[2] | 27.7 ± 2.0[2] | 0.98 |
| Stress (1 h HLL) | 31[2] | 26.8 ± 2.1[2] | 38.0 ± 3.6 | 61.5 ± 6.6 | 0.66 |
| Ethanol and stress | 81[2] | 45.7 ± 5.9[2] | 22.3 ± 1.9[2] | 26.9 ± 2.2[2] | 0.83 |

[1] μg/g, mean ± SEM of 6 rats.
[2] $p < 0.05$.

*Table III.* Effect of alcohol and stress on the uptake, distribution and elimination of hexobarbital in rats [from ref. 3]

| Treatment[1] | Peak concentration | TV 1/2[2] | AUC[3] |
|---|---|---|---|
| Control | 37 | 120 | 4,067 |
| Ethanol | 27 | 225 | 2,908 |
| Stress | 23 | 240 | 3,268 |
| Ethanol and stress | 45 | 360 | 4,840 |

[1] See tables I and II.
[2] The elimination phase only.
[3] AUC = area under curve.

did not alter in vitro metabolism of hexobarbital significantly in controls. Adrenalectomy blocked the inhibition of hexobarbital hydroxylase by alcohol and HLL alone and when HLL and alcohol were combined.

This series of experiments showed that in vitro hexobarbital metabolism was inhibited after acute pretreatment with alcohol and acute stress (HLL). Combining the stress and alcohol pretreatments produced an additive effect on inhibition of hexobarbital hydroxylase (table II). Duration of 'sleep time' (loss of righting ability after hexobarbital administration) was not an accurate measure of disappearance of hexobarbital because the brain sensitivity and the distribution of hexobarbital are both altered by the pretreatments

*Table IV.* Effect of alcohol and stress on plasma corticosterone levels and hexobarbital hydroxylase activity in normal adrenalectomized rats [from ref. 2]

| Treatment | Intact | | Adrenalectomized | |
|---|---|---|---|---|
| | corticosterone[1] | hexobarbital[2] | corticosterone | hexobarbital |
| Control | 8.9 ± 2.8 | 6.69 ± 0.6 | 1.1 ± 0.6 | 6.10 ± 0.44 |
| Ethanol | 26.1 ± 5.2[3] | 4.59 ± 0.07[3] | 0.9 ± 0.6 | 6.12 ± 0.41 |
| Stress | 18.9 ± 2.2[3] | 3.66 ± 0.17[3] | 0.5 ± 0.4 | 6.10 ± 0.30 |
| Ethanol and stress | 46.8 ± 4.5[3] | 1.27 ± 0.04[3] | 0.4 ± 0.4 | 6.17 ± 0.45 |

[1] Corticosterone values are expressed in μg/100 ml plasma as means ± SEM in 6–8 rats.
[2] Hexobarbital hydroxylase values are expressed in μmol hexobarbital metabolized/g liver/10 min as means ± SEM in 6–8 rats.
[3] $p < 0.05$.

(table II, III). The half time for elimination of hexobarbital from the plasma was correlated with the in vitro liver metabolism (table II). The decreases in microsomal P450 levels also correlated with decreases in hexobarbital metabolism in vitro.

Plasma corticosterone levels also correlated well with the inhibition of hexobarbital hydroxylase activity in vitro (table IV). Alcohol as well as HLL stress increased plasma corticosterone levels 2- to 3-fold, respectively, and the combination produced a 5-fold increase in plasma corticosterone. Adrenalectomy blocked the effects of alcohol, as well as the effects of stress, on metabolism of hexobarbital. These results suggested that the inhibition could be mediated through corticosterone or another adrenal hormone. However, it was found that there was no inhibitor present in the cytosol of treated rats (data not shown).

Current studies are designed to extend these investigations to other MFO enzymes and to study the action on environmental toxins. The effect of alcohol and stress on the N-demethylase pathway of aminopyrine, a type I substrate, are shown in table V. The pattern of inhibition of aminopyrine N-demethylase parallels that found in the hexobarbital studies.

The effect of alcohol and stress on aniline hydroxylase, a type II substrate, is shown in table VI which also shows the effect of alcohol and stress in a type II substrate metabolism in the liver.

The metabolism of BAP also involves MFO pathways in the liver. The effects of the alcohol pretreatment and corticosterone, a hormone released

*Table V.* Effect of alcohol and stress on aminopyrine N-demethylase activity in 9,000 *g* supernate fractions of rat liver [from ref. 1]

| Treatment | Aminopyrine N-demethylase activity[1] | % inhibition |
|---|---|---|
| Control | 680 ± 49 | 0 |
| Alcohol (3 g/kg) | 500 ± 40[2] | 26 |
| Stress (1 h HLL) | 490 ± 36[2] | 28 |
| Alcohol and stress | 370 ± 27[2] | 48 |

[1] nmol of formaldehyde formed/30 min/250 mg liver as means ± SEM in 5 rats.
[2] $p < 0.05$.

*Table VI.* Effect of alcohol and stress on aniline hydroxylase activity of 9,000 *g* supernatant fractions of rat liver [from ref. 1]

| Treatment | Aniline hydroxylase[1] | % inhibition |
|---|---|---|
| Control | 183 ± 21 | 0 |
| Alcohol (2 g/kg) | 157 ± 14[2] | 14 |
| Stress (1/2 h HLL) | 152 ± 18[2] | 17 |
| Alcohol and stress | 136 ± 8[2] | 26 |

[1] nmol of *p*-hydroxyaminophenol formed/30 min/250 mg tissue as means ± SEM in 5 rats.
[2] $p < 0.05$.

*Table VII.* Effect of acute ethanol and corticosterone on BAP disappearance in microsomal preparations of rats administered 3-MC (i.p.) 3 mg/kg, 40 h before treatment [from ref. 1]

| Treatment | n | BAP disappearance[1] | % inhibition |
|---|---|---|---|
| Controls | 10 | 5.24 ± 0.67 | 0 |
| Alcohol (3 g/kg) | 5 | 4.92 ± 0.98 | 6 |
| Alcohol (5 g/kg) | 8 | 3.45 ± 0.46[2] | 34 |
| Corticosterone (12.5 mg/kg) | 8 | 2.45 ± 0.37[2] | 53 |

[1] Values expressed as nmol BAP/min/mg protein ± SEM in (n) rats.
[2] $p < 0.05$.

by the adrenal under stress conditions, on BAP disappearance are shown in table VII. Thus, the effects of acute alcohol and stress produce a pattern of inhibition of microsomal metabolism of aminopyrine, aniline and BAP similar to hexobarbital.

*Effect of Adrenalectomy on the Alcohol-Stress MFO Inhibition*

The effect of adrenalectomy on the alcohol and corticosterone inhibition of microsomal metabolism of BAP, aminopyrine and aniline was determined. No inhibition was produced by either alcohol (5 mg/kg) or corticosterone administration on any of the three substrates investigated in adrenalectomized rats. Blood alcohol concentrations, determined at 1 h, were found to be 0.17 ± 0.01 and 0.29 ± 0.01 g% w/v after 3- and 5-kg doses, respectively. Stress did not alter the blood alcohol concentration at 1 h.

## *Discussion*

Acute alcohol exposure and acute stress (1 h) inhibit three important MFO pathways of drug metabolism: hexobarbital hydroxylase, aminopyrine-N-demethylase and aniline hydroxylase (tables I–VII). Combining the alcohol and stress produced an additive inhibition of the MFO enzymes as well as an additive increase in the plasma levels of corticosterone. Adrenalectomy blocks the MFO inhibition produced by alcohol, stress and corticosterone. In vitro disappearance of BAP is inhibited by acute alcohol and corticosterone and, similar to the other MFO enzymes, it is blocked by adrenalectomy. Inhibition of BAP metabolism is produced at clinically relevant blood alcohol levels and short-time exposures (30 min to 1 h).

The effect of alcohol on drug metabolism is a biphasic phenomenon in which inhibition occurs immediately after exposure, followed by stimulation within 3 h [12, 14, 16]. The findings in our laboratories using acute alcohol treatment compare favorably with these reports. Acute stress inhibition of MFO is similar to that of acute alcohol. The observation that the effects of stress and corticosterone are additive with those of alcohol are consistent with this and suggest that they are acting through a similar mechanism. It is possible that corticosterone is the common factor. However, the observation that corticosterone itself does not inhibit MFO activity in adrenalectomized rats is inconsistent with this hypothesis. The research does show that alcohol as well as stress requires an intact adrenal for the inhibitory effect on liver metabolism to occur.

## *References*


1 Brown, D.R.; Ning, J.; Bogdanffy, M.S.; Krull, I.; Kuttab, S.; Shargel, L.: Alcohol-stress inhibition of mixed function oxidase metabolism in the rat. Toxicologist *2:* 129 (1982).

2 Chung, H.; Brown, D.R.: The mechanism of the effect of acute stress on hexobarbital metabolism. Toxicol. appl. Pharmacol. *37:* 313–318 (1976).

3 Chung, H.; Brown, D.R.: Alcohol-hexobarbital interaction in rats under acute stress. Life Sci. *18:* 123–128 (1976).

4 Chung, H.; Brown, D.R.: Mechanism of the effect of acute ethanol on hexobarbital metabolism. Biochem. Pharmac. *25:* 1613–1616 (1976).

5 Chung, H.; Brown, D.R.: The mechanism of the effect of alcohol-stress in combination on the biotransformation of hexobarbital; Fouts, Gut, Industrial and environmental xenobiotics, pp. 284–286 (Excerpta Medica, Amsterdam 1978).

6 Conney, A.H.; Burns, J.J.: Factors influencing drug metabolism. Adv. Pharmacol. *1:* 31–58 (1962).

7 Cooper, J.R.; Brodie, B.B.: The enzymatic metabolism of hexobarbital (Evipal). J. Pharmac. exp. Ther. *114:* 409–417 (1955).

8 Kato, R.; Gilette, J.: Sex differences in the effects of abnormal physiological states on the metabolism of drugs by rat liver microsomes. J. Pharmac. exp. Ther. *150:* 285–291 (1965).

9 Kinoshita, N.; Shears, B.; Gelboin, H.V.: K-region and non-K-region metabolism of benzo(a)pyrene by rat liver microsomes. Cancer Res. *33:* 1937–1944 (1973).

10 Lieber, C.S.; Seitz, H.K.; Garro, A.J.; Warner, T.M.: Alcohol-related diseases and carcinogenesis. Cancer Res. *39:* 2863–2886 (1979).

11 McMahon, R.E.; Culp, H.W.; Mills, J.; Marshall, F.J.: Demethylation studies IV. The in vitro and in vivo cleavage of alkyl- and aryl alkyl-*p*-nitrophenyl ethers. J. med. Chem. *6:* 343–346 (1963).

12 Pennington, S.N.; Smith, C.F.; Tapscott, B.: Effect of an acute dose of ethanol on rat liver microsomal mixed function oxygenase components and membrane lipid composition. Alcoholism exp. Res. *2:* 311–316 (1978).

13 Sellers, E.M.; Holloway, M.R.: Drug kinetics and alcohol ingestion. Clin. Pharmacokinet. *3:* 440–452 (1978).

14 Sotaniemi, E.A.; Anttila, M.; Rautio, A.; Stengard, J.; Saukko, P.; Jarvensivu, P.: Propanolol and sotalol metabolism after a drinking party. Clin. Pharmacol. Ther. *29:* 705–710 (1981).

15 Soyka, L.F.: Determinants of hepatic aminopyrine demethylase activity. Biochem. Pharmac. *18:* 1029 1038 (1969).

16 Rubin, C.; Lieber, C.S.: Hepatic microsomal enzymes in man and rat: induction and inhibition by ethanol. Science *162:* 690–691 (1968).

17 Zenker, N.; Berstein, D.E.: The estimation of small amount of corticosterone in rat plasma. J. biol. Chem. *231:* 695–701 (1958).


Dr. D.R. Brown, Northeastern University, DRB 312 MV, College of Pharmacy and Allied Health Professions, 360 Huntington Avenue, Boston, MA 02115 (USA)

# Method Development for Safety Evaluation

Safety Evaluation and Regulation of Chemicals. 1st Int. Conf., Boston, Mass., 1982, pp. 191–192 (Karger, Basel 1983)

## Introduction to Method Development Session

*F. Homburger*

Bio-Research Institute, Inc., Cambridge, Mass., USA

The past few years have brought about a revolution in the methodology of safety evaluation of chemicals. The cumbersome but relatively simple straightforward lifetime feeding study has become, through compliance with GLPs, a nightmare for all concerned, from those who conceive the studies and pay for them to those who carry them out and those who finally interpret them. Completion of a study and submission of a report does not end this disquieting dream-like situation, for archivation remains, and quite likely FDA inspections will follow, which will dig back into the maze of accumulated data, SOPs, protocols and amendments.

This new complexity of the evaluation procedure together with its often confusing results have proven to be strong stimuli towards abbreviated tests and simpler procedures. Added to this are the trends towards protecting animals, even talk of 'animal rights' and efforts to reduce the number of animals in such routine procedures as the $LD_{50}$.

Independent of these practical and political considerations, there has been an explosion of fundamental knowledge that may eventually make the conventional evaluation procedure obsolete and unnecessary. This theme will be developed by Dr. *Golberg*. Dr. *Williams* will discuss a new approach which tends to minimize in vivo testing and is based on this very same basic knowledge.

For the immediate future the toxicologist must continue to rely on animal bioassay for the ultimate determination of safety and risk, at least until legislation will make it possible to do without such bioassays. It is therefore important to sharpen the existing procedures to a point of refinement which will yield reliable results more quickly and more economically than is now possible. Dr. *Ward* will examine methods of evaluating the performance of animal populations in long-term tests, Dr. *Reznik-Schüller* will report novel means for the morphological detection of early lesions in animals and I shall talk about refining carcinogenesis bioassay by the use of a species with low spontaneous tumor incidence and high susceptibility to administered carcinogens, the Syrian hamster. I shall also touch on some of our recent work in cooperation with American Science and Engineering, Inc., of Cambridge, suggesting that application to tumor diagnosis of the X-ray method known as computer-assisted tomography in rodent tests may contribute to shortening the time required for in vivo carcinogenesis bioassay. Dr. *Henry* will report improved inhalation exposure methods, Dr. *Balazs* will review some rare reactions of animals to toxic substances and their early detection and Dr. *Mazué* will demonstrate that blood and tissue enzymatic tumor markers can be used to advantage to detect early changes in test animals.

F. Homburger, MD, Bio-Research Institute, Inc., 9 Commercial Avenue, Cambridge, MA 02141 (USA)

Safety Evaluation and Regulation of Chemicals. 1st Int. Conf., Boston, Mass., 1982, pp. 193–199 (Karger, Basel 1983)

# The Next 25 Years in Toxicology

*Leon Golberg*

Division of Community/Occupational Medicine, Duke University Medical Center, Durham, N.C., USA

I find myself a strange bedfellow among the futurists. For the most part they are doomsayers – whether one is mesmerized by the brilliance of *George Orwell's* vision of *1984,* or oppressed by *Theodore Roszak's* anticipation of 'Environmental *collapse, world poverty, technocratic elitism, psychic alienation, the death of the soul…'.* As far as toxicology can influence the environmental and human condition, I find much cause for optimism; in fact, I feel that there has never been a time of such ferment and so much promise. Of course, as *Halley and Vatter* [9] point out, at any given moment, we can only draw 'inferences from specifics contained in today's most advanced innovations' xut the current state of these innovations clearly indicates that we can expect profound changes in the immediate future of toxicology.

## *Present Weaknesses*

A half-century of scientific advances has provided sufficient information to make possible a clearer realization of the gaps in our overall knowledge of the phenomena that pertain to toxicology [5]. It is these gaps that render progress so difficult beyond the purely observational aspects of our science to the necessary interpretative insights based on the awareness of mechanisms of toxic action in man and animals. The spectrum of changes, in relation to increasing dosage and/or duration of exposure, ranges from physiological adaptation to the limits of homeostatic capacity; and, once those limits are exceeded, the effects extend to pathological (structural and functional) derangements [11]. These lesions may prove lethal, or severe enough to cause permanent injury. Any damage short of acute lethality is usually accompanied by some evidence of attempts at repair and restoration of structure or function, or both [6].

All this has been known for many years: yet to what extent have the mechanisms of these processes been elucidated? Even such intensively studied phenomena as inflammation are only now beginning to yield up their secrets, thanks to the application of new technology. Greater ignorance prevails in most areas of homeostasis, in the understanding of defensive mechanisms throughout the body and of the limits of their capacity. Above all, fundamental data are lacking concerning the species, strain (or genetic background), sex, age, disease, diet and lifestyle determinants that influence both the limits of homeostatic response and the nature and course of toxic events that ensue when such limits are exceeded [10]. In this context, neuroendocrine control mechanisms and immune responses are particularly important.

Finally, research has focussed almost exclusively on individual exposure, for the most part neglecting the complexities of what is a fact of life, namely the multiplicity of exposures in the workplace, the home and the environment. Drug interactions and, to a lesser extent, pesticide interactions have attracted attention; but no general methodology has emerged for the study of the great majority of the additive, synergistic or antagonistic actions manifested by multiple exposures to chemical and physical agents. In fact, the conditions that make life possible – such as sunlight and oxygen – not only have toxic potentialities in themselves but also interact with exposure to other agents to bring about phototoxic and other adverse effects.

### *The Overall Prospect*

Two overlapping phases of development will be proceeding simultaneously during the coming decades. Phase 1 involves further elaboration and perfection of existing techniques in toxicology. There is certainly much room for improvement of animal diets, breeding, care and data handling and processing. Automation of labor-intensive operations will proceed apace. More effort will be devoted to the definition of preneoplastic states by histochemical and other means, as well as the establishment of objective criteria for the diagnosis and classification of tumors in laboratory animals. Phase 2, which has only just begun, involves the application of new technology to achieve ends that have proved to be beyond the capacity of the traditional methods, or have been so tedious and difficult that progress has slowed virtually to a standstill. Included in this category are such objectives as dose monitoring in tissues, especially where covalent binding to target macromolecules has taken place within the cell. Such measurements are now becoming possible at levels of increasing sensitivity, as well as at a high degree of specificity. The use of immunological techniques for this purpose has culminated in the application of monoclonal antibodies to follow the course of DNA adduct formation and excision at a reputed sensitivity of one such adduct per cell [12].

The impact of the new techniques of phase 2 on the developments of phase 1 has been evident for some time. The introduction by *Ames* et al. [1] of genetically modified strains of *Salmonella typhimurium*, coupled with the use of an in vitro activating system, has revolutionized the study of genotoxicity. Others have followed in these footsteps. The development of an automated system for karyotyping and quantifying chromosomal aberrations still eludes us, despite many efforts in that direction. The enumeration of sister chromatid exchanges (SCE) as a substitute opens up new questions such as the significance of SCE and the range of normal background incidence in the population, and among special groups such as smokers [3].

### *From in vivo to in vitro*

Of the three levels of biological organization, the current trend is to start with the whole animal and to progress to cellular and molecular levels. For purposes of extrapolation, correlative and integrative efforts have to move in the opposite direction. Ideally, a hierarchical approach is preferable – one that provides full understanding on the basis of interdisciplinary competence. It should begin with the study of the implications of structure, chemical and physical properties, especially chemical reactivity in activated form at the lowest levels of biological organization, and proceed to progressively more complex systems, building at each stage upon the knowledge gathered at the previous step.

A combination of circumstances is conspiring to force a trend away from the use of whole animals in toxicology. The activities of animals' rights groups, munificent provision of funds for research on substitutes for animal tests, and the increasing sophistication and scope of *in vitro* approaches are bound to have a considerable impact. An example is the use of single cell types or of separated component parts studied in isolation. Centrifugal elutriation techniques have been used to separate the individual cell types present in the liver, lung and kidney. Cloned human and murine cell lines, such as Leydig and Sertoli cells, represent a departure beyond mere separation of specific cell types. Microdissection of specific sites in the brain has permitted the separation and subsequent purification of single nerve cells and other component parts that have been studied in isolation: cell bodies, glia, myelin, nuclei, synaptosomes and synaptic vesicles are some examples. Microchemical methods permit measurement of neurotransmitters in single neurons ($10^{-12}$–$10^{-15}$ moles/cell). Microtechnology, microinjection of single cells (for instance, in tumor DNA transfection experiments) microdensitometry, microfluorimetry and immunofluorescence techniques have been joined by the laser microprobe mass analyzer for micro mass spectrometry.

## *The Proper Study of Mankind*

Much of the current effort in toxicology is misdirected. In times past, experimental work largely employed animal models as surrogates for man, even when people were already being exposed to the particular chemical in question and would have been more logical objects for study of toxic effects. Mounting difficulties in the investigation of such individuals, and even more so in the use of human volunteers, have militated against major advances in human toxicology.

An even more serious obstacle has been the lack of noninvasive methods possessing sufficient sensitivity and specificity that would reflect overall exposure (both severity and duration), and particularly the target dose reaching specific organs. The need for biological dose monitors has been repeatedly emphasized [8], and the desirable characteristics of an ideal procedure have been listed [15, 18]. High on the list is the need for information on potential health risks of the observed target doses. Sampling for indices of exposure and effect has included breath, urine, sweat, saliva, mucus, semen, blood, hair, nails, biopsy specimens of skin, fat, liver, nasal mucosa and tissues obtained during surgery or necropsy.

The need to ascertain the health effects of inadvertent exposure to pinpoint sources of pollution has become urgent [7]. Recent developments have been notable for the progress made in noninvasive, nondestructive techniques for human and animal studies [16]. The use of nonradioactive isotopes as labels has facilitated human metabolic studies. Immunological techniques, and particularly monoclonal antibodies, were mentioned above. Noninvasive imaging methods such as CAT scanning of man and experimental animals, positron emission tomography and NMR imaging [2] offer largely unexplored potentialities for the toxicologist [4].

An important technical advance has been the introduction of a technique for readily quantifying SCEs in human and animal chromosomes [3], especially in peripheral lymphocytes. As a further step, attention has now been directed to the so-called 'high-frequency cells' (HFCs) which will perhaps prove to be more persistent evidence of DNA damage long after exposure to a mutagen has ceased [17]. At the very least, the HFCs should provide a means of distinguishing more clearly than at present between the normal background incidence of SCEs and the effects of mutagens or carcinogens.

Try to imagine the benefits to be derived from the availability of methods of detecting and quantifying the health effects of exposure to the current environmental or dietary levels of mutagens and carcinogens! Consider the information that will flow from studies on the health effects of common foods and beverages, in relation to their interactions with the genetic background, lifestyle and other characteristics of the individual consumer! The gross clinical manifestations that accompanied past occupational exposures to chemicals have largely disappeared with increasingly tight control of levels in the workplace. Does the absence of readily recognizable disease equate with health and safety? This question can only be answered by applying new techniques possessing correspondingly high sensitivity and specificity. Some of these new methods will undoubtedly be introduced – much as SCEs have been – as a result of advances in the basic sciences. By keeping in close touch with such progress, notably in the field of biotechnology, the toxicologist will be ready to seize the opportunities that appear on the horizon.

## *The Challenge of the Future*

The emphasis in toxicology has hitherto been directed mainly to normal, healthy animals as models of the totality of the human population. Diseased and disabled people, including children, have always been out there, with little attention paid to their special susceptibilities. Within the next 25 years, the toxicologist will be serving a very different society. The numbers and proportion of aged and disabled individuals are expected to grow in the developed countries, but even more so in the developing countries. With an anticipated rise in the world population of 2 thousand million people in the next 20–30 years, the number of disabled persons is expected to increase dramatically [19].

Much of existing and anticipated disease and disability is associated with the impact of food and nutrition on health. Accordingly, the attention of all facets of research in preventive medicine will have to be devoted to this aspect, paying particular attention to the interactions with one another of nutritional deficiency, nutrient imbalance and toxicants in raw and cooked food, as well as their combined interactions with environmental pollutants [13].

The only hope of feeding the countless millions of the earth's inhabitants will lie in the products of the new technologies. Not only new processes, but a host of new foods and food components (and, alas, contaminants) will challenge the resources and ingenuity of the toxicologist. Such products often present formidable difficulties in safety evaluation, as seen in the past with irradiated foods, single cell protein for human consumption, and fish protein concentrate [14].

The new foods and processes will represent but one aspect of the application of biotechnology to yield new generations of therapeutic agents, human and animal vaccines, agricultural chemicals and industrial chemical building-blocks. The latter will be based increasingly on feedstocks derived from biomass and other sources unrelated to fossil fuels. Chemical intermediates that are curiosities or even unknown at present will come flooding in, with bewildering variety and complexity.

Scientifically, the crippling preoccupation of toxicologists with cancer hazards will have given way to a broader and more balanced outlook. In the first place, means of accurately predicting and assessing cancer risk will have vastly improved. Secondly, the means available for appraising factors that initiate or modulate the development of cancer in man and animals will have made it possible to draw clear distinctions between various categories of hazard. Thirdly, society will place much greater emphasis, than it does at present, on degenerative diseases and other aging phenomena in relation to toxic exposures. All these influences will serve to direct toxicologists' activities towards more effective provision of data for preventive medicine. In the final analysis, the effectiveness of the application of this new knowledge will depend on public understanding and receptivity to new ideas. The toxicologist has a role to play in this area also.

## *References*

1 Ames, B.N.; Durston, W.E.; Yamasaki, E.; Lee, F.D.: Carcinogens are mutagens: a simple test system combining liver homogenates for activation and bacteria for detection. Proc. natn. Acad. Sci. USA *70:* 2281–2285 (1973).

2 Brownell, G.L.; Budinger, T.F.; Lauterbur, P.C.; McGeer, P.L.: Positron tomography and nuclear magnetic resonance imaging. Science *215:* 619–626 (1982).

3 Carrano, A.V.; Thompson, L.H.: Sister chromatid exchange; in Wolff, Sister chromatid exchange (Wiley, New York 1982).

4 Gadian, D.G.: Metabolic studies of whole animals and humans using phosphorus nuclear magnetic resonance. Biosci. Rep. *1:* 449–460 (1981).

5 Golberg, L.: Toxicology: has a new era dawned? Pharmac. Rev. *30:* 351–370 (1979).

6 Golberg, L.: The dangers of new discoveries and the discovery of new dangers; in Human health and environmental toxicants. R. Soc. Med. Int. Congr. Symp. Ser., No. 17, pp. 19–39 (1979).

7 Goldstein, I.F.: The use of biological markers in studies of health effects of pollutants. Environ. Res. *25:* 236–240 (1981).

8 Gompertz, D.: Assessment of risk by biological monitoring. Br. J. ind. Med. *38:* 198–201 (1981).

9 Halley, R.B.; Vatter, H.G.: Technology and the future as history: a critical review of futurism. Technol. Cult. *19:* 53–82 (1978).

10 Harris, C.C.; Mulvihill, J.J.; Thorgeirsson, S.S.; Minna, J.D.: Individual differences in cancer susceptibility. Ann. int. Med. *92:* 809–825 (1980).

11 Hatch, T.F.: Changing objectives in occupational health. Am. ind. Hyg. Ass. J. *23:* 1–7 (1962).

12 Haugen, A.; Groopman, L.D.; Hsu, I.-C.; Goodrich, G.R.; Wogan, G.N.; Harris, C.C.: Monoclonal antibody to aflatoxin $B_1$-modified DNA detected by enzyme immunoassay. Proc. natn. Acad. Sci. USA *78:* 4124–4217 (1981).

13 Hook, E.B.: Human teratogenic and mutagenic markers in monitoring about point sources of pollution. Environ. Res. *25:* 178–203 (1981).

14 Friedman, L.; Glaser, O.G.; Brown, N.L.; Pariser, E.R.: The wholesomeness of fish protein concentrate: a new approach to the evaluation of food safety. Toxicol. appl. Pharmacol. *18:* 239–252 (1971).

15 Lauwerys, R.: Biological criteria for selected industrial toxic chemicals: a review. Scand. J. Work environ. Hlth *1:* 139–172 (1975).

16 Maugh, T. H. II: Biological markers for chemical exposure. Science *215:* 643–647 (1982).

17 Stetka, D.G.; Minkler, J.; Carrano, A.V.: Induction of long-lived chromosome damage, as manifested by sister chromatid exchange, in lymphocytes of animals exposed to mitomycin-C. Mutat. Res. *51:* 383–396 (1978).

18 World Health Organization: Early detection of health impairment in occupational exposure to health hazards. Tech. Rep. Ser. Wld Hlth Org., No. 571 (1975).

19 World Health Organization: Disability prevention and rehabilitation. Tech. Rep. Ser. Wld Hlth Org., No. 668 (1981).

Dr. L. Golberg, Division of Community/Occupational Medicine, Duke University Medical Center, Box 2914, Durham, NC 27710 (USA)

Safety Evaluation and Regulation of Chemicals. 1st Int. Conf., Boston, Mass., 1982, pp. 200–209 (Karger, Basel 1983)

# New Approaches to Carcinogen Bioassay

*Gary M. Williams, John H. Weisburger*

Naylor Dana Institute for Disease Prevention, American Health Foundation, Valhalla, N.Y., USA

## *Carcinogen Testing*

The usual approach to assessing the carcinogenicity of a chemical has been chronic administration to experimental animals. Currently, the standard typical bioassay for the detection of chemical carcinogens involves administration of the maximally tolerated dose and one half that dose to groups of 50 male and female rats and mice, and occasionally hamsters, over an approximately 2-year interval. The difficulties with this approach, which have been reviewed by *Weisburger and Williams* [9], include the unnecessary use of animals, substantial costs, and the requirement for scarce specialty skills. Moreover, an interval of up to 5 years can be required for the delineation of carcinogenicity, which represents an unacceptable period of human exposure for a chemical already in use.

Chronic bioassay techniques were elaborated and standardized in the 1960s, before their limitations were apparent and before the mechanisms of carcinogenesis were adequately understood. Thus, the tests conducted by the National Cancer Institute and other agencies have frequently yielded ambiguous results [9]. Moreover, in light of current knowledge, results of such tests in the absence of collateral information, can no longer be accepted as a definitive indication of potential human hazard.

## *Chemical Carcinogenesis*

A major class of carcinogens designated as pro- or precarcinogens, or indirectly acting carcinogens, requires conversion by host enzyme systems to a more active agent, the proximate carcinogen, which in turn is further metabolized to the ultimate carcinogen [10]. The latter agent can interact directly with cellular macromolecules initiating a sequence of steps leading to cancer. Synthetic, directly acting carcinogens have such properties inherent in their structure. The ultimate carcinogens of this type were originally shown by *Miller and Miller* [4] to be electrophilic reactants, which possess a positive charge such that they can react covalently with nucleophilic components in cells. The genetic material, DNA, has emerged as potentially the most significant target for such reactions [1].

The ability of some carcinogens to damage DNA has been referred to as genotoxicity [11]. For such chemicals, the carcinogenic process appears to begin with the interaction between the ultimate carcinogenic form of the genotoxin with DNA. In a quiescent nonduplicating cell, an opportunity exists for this damage to be repaired. However, duplication of a cell with modified DNA can result in mispairing of bases or transposition of sequences and thereby give rise to progeny with an abnormal genome corresponding to what is commonly called the dormant or latent tumor cell. Under permissive conditions, such an abnormal cell can proliferate and give rise to a neoplasm.

The ability of certain carcinogens to interact with DNA has led to an association between carcinogenicity and mutagenicity, but some definite exceptions to this correlation exits. A number of important environmental chemicals such as DDT and other chlorinated hydrocarbons, hormones and drugs, such as phenobarbital, caused tumors in standard animal bioassays, yet were uniformly inactive in mutagenicity tests. By the operational definition of carcinogens, all such products would be called carcinogens since they increase cancer in experimental animals. Yet, just as the concept of electrophilic reactants has provided a unifying understanding for the action of structural types of organic carcinogens, additional mechanistic concepts will have to be applied to these other agents. In contrast to genotoxic carcinogens, these agents, lacking the capacity to damage DNA, exert their carcinogenic effects through incompletely known mechanisms which have been called epigenetic effects [11].

These concepts on the mechanisms of different types of agents have been translated by *Weisburger and Williams* [11] into two broad categories of carcinogen classification, genotoxic and epigenetic, which have been further subdivided into eight classes. The recognition of multiple modes of action for carcinogens has major implications for the design of test protocols and for the interpretation of results. In an attempt to incorporate this new knowledge of chemical carcinogenesis into a systematic approach for carcinogen bioassay, a decision point approach to delineating possible human carcinogenic and mutagenic risks was proposed in 1978 [8].

### *Decision Point Approach*

The decision point approach consists of five sequential steps in the evaluation of the potential carcinogenicity of chemicals. This approach was formulated to incorporate into chemical evaluation several new concepts in chemical carcinogenesis. Of prime importance among these is the concept that chemicals could produce an increase in the tumor incidence in exposed animals by several distinct mechanisms, which could be categorized as either genotoxic or epigenetic. The decision point approach takes into account these two categories of carcinogens in several ways: (a) a battery of short-term tests is implemented in an effort to provide test systems to identify genotoxic carcinogens at an early stage of testing, and which may also respond to epigenetic agents, and (b) it is recognized that all forms of subchronic testing may fail to detect nongenotoxic chemicals that can induce neoplasms in animals under specific conditions of chronic exposure.

The decision point approach is a systematic approach to the determination of the potential carcinogenicity of chemicals which provides a framework in which to minimize the necessary testing in chemical evaluation, and at the same time develops an understanding of the mechanism of action of a test chemical. An outline of the approach is given in table I. It involves a systematic step-wise progression of tests. A critical evaluation of information obtained and its significance in relation to the testing objective is performed at the end of each phase. A decision is made as to whether the data available are sufficient to reach a definitive conclusion or whether a higher level of tests is required. Attention is paid to both qualitative and quantitative effects.

#### *Stage A: Structure of Chemical*

The evaluation begins with a consideration of the structure of the chemical. Among the reasons for this is the important fact that predictions as to whether or not a given chemical might be carcinogenic can be made with fair success within certain classes of chemicals. This is particularly true in the case of chemicals of a type that includes known genotoxic carcinogens. However, structure must always be considered in relationship to species parameters of biotransformation which underlie species selectivity to many activation-dependent carcinogens [10].

*Table I.* Decision point approach in carcinogen testing

| | |
|---|---|
| A | Structure of chemical |
| B | In vitro short-term tests |
| 1 | DNA repair |
| 2 | Bacterial mutagenesis |
| 3 | Mammalian mutagenesis |
| 4 | Chromosome tests |
| (5) | Cell transformation[a] |
| *Decision Point 1:* All tests under A and B | |
| C | Limited in vivo bioassays |
| 1 | Skin tumor induction in mice |
| 2 | Pulmonary tumor induction in mice |
| 3 | Breast cancer induction in female Sprague-Dawley rats |
| 4 | Altered foci induction in rodent liver |
| 5 | Assays for promoters |
| *Decision Point 2:* Evaluation from the beginning plus results from any of the appropriate tests under C | |
| D | Chronic bioassay |
| E | Final evaluation |
| Evaluation of all results available that must include data from A and B to provide basis for mechanistic considerations | |

[a] Included only if data from other tests inconclusive.

Information on structure and metabolism also provides a guide to the selection among limited bioassays at stage C and, as more information from research accrues, may eventually contribute to selection of specific short-term tests at stage B.

*Stage B: Battery of Short-Term Tests*

No individual test that has been studied adequately has detected all carcinogens tested and thus, from this practical consideration alone, a battery of tests is essential. However, the importance of a battery becomes obvious upon consideration of the complexities of biotransformation and mechanisms of action of chemical carcinogens.

Over a hundred short-term tests for carcinogens and mutagens are available [2]. From among these, a variety of batteries have been proposed [*Williams* in 13]. It is not really feasible to compare in all regards the utility of different batteries. In the decision point approach four primary tests and one supplemental test were selected on the basis of the reliability and biologic significance of their end points, the nature of their biotransformation systems, and their availability. Critical appraisals of all these systems have been conducted [3]. The tests selected are as follows.

*1. DNA Repair.* Of the systems available, the hepatocyte primary culture/DNA repair test developed by *Williams* [in 6, 13] has demonstrated substantial sensitivity and reliability with activation-dependent procarcinogens, including some not readily detected in other systems. It is one of the few tests that provides intact cell metabolism.

*2. Bacterial Mutagenesis.* Because of the extensive data base and good correlation with carcinogenicity, the Ames test, measuring reversion of tester strains of mutant *Salmonella typhimurium* from a histidine requirement to prototrophy [*Haroun and Ames* in 6] in its recent versions, including liquid phase incubations, is recommended for a bacterial mutagenesis test.

*3. Mammalian Mutagenesis.* Mutation at the hypoxanthine-guanine phosphoribosyltransferase locus in mammalian cells is a well-characterized mutational system which can readily be measured by conversion of cells to resistance to toxic purine analogs. The V79 and CHO lines are two cell lines used in purine analog resistance assays that have been extensively validated [3, *Hsie* in 13]. These fibroblast-like cells possess little ability to activate carcinogens and this deficiency must be met by supplementation with either cocultivated cells or enzyme preparations.

*4. Chromosome Test.* Because of simplicity and objectivity, sister chromatid exchange is recommended for a test of chromosome damage. The best validated system at present is that using CHO cells [*Wolff* in 6].

*5. Cell Transformation.* Most transformation assays involve fibroblasts and measure a morphological alteration in the cells. Assays for changes in growth properties related to neoplasia, such as growth in soft agar [*Styles* in 5] and incorporation of more relevant cell types such as epithelial cells [*Borek* in 5] are desirable. The systems of *Pienta* [in 6, 13] and of *Heidelberger* [in 5] appear to be sufficiently widely used to be considered as potential supplements to the other four tests if deemed necessary.

This battery has been determined to be highly efficient in detecting most types of genotoxic carcinogens [12].

*Decision Point 1*

The six steps (A plus B, 1–5) discussed thus far are the basis for preliminary decision making with a high degree of sensitivity and specificity for genotoxic carcinogens and mutagens.

If definite evidence of genotoxicity in more than one test has been obtained, a chemical is highly suspect. In particular, because of their complementary nature, positive results in the test systems of *Ames* and of *Williams* provide a virtually certain indication of carcinogenic potential. An agent that is DNA damaging, mutagenic, and produces chromosome effects is likewise virtually certain to be carcinogenic and, regardless, represents an unequivocal toxic hazard.

On the other hand, genotoxicity found in only one test requires interpretation with caution. For example, several types of chemicals such as intercalating agents are mutagenic to bacteria but not reliably carcinogenic. Positive results have also been seen in bacteria using synthetic phenolic compounds or natural products with phenolic structures such as flavones. In vivo, such compounds are conjugated and excreted readily. The subcellular enzyme preparations which are required in many in vitro systems are deficient in certain detoxification reactions and thus may yield an inaccurate indication of carcinogenic potential.

Any positive results in the in vitro battery can be extended through limited in vivo bioassays (stage C), without the need to conduct a full scale costly and time-consuming chronic bioassay. If all the preceding test systems yield no indication of genotoxicity, however, the priority for further testing depends on two criteria: (1) the structure and known physiological properties (e.g. hormone) of the material, and (2) the potential human exposure. If substantial human exposure is likely, careful consideration should be given to undertaking additional testing. The chemical structure and the properties of the material can provide guidance on the proper course of action.

*Stage C. Limited in vivo Bioassays*

This stage of evaluation employs tests that will provide further evidence of potential hazard of chemicals positive for genotoxicity without the necessity of undertaking chronic bioassay.

The in vivo tests recommended at this stage are those that will provide definitive evidence of carcinogenicity, including cocarcinogenicity and promotion, in a relatively short period (i.e. 30 weeks or less). Unlike the in vitro tests, these are not applied as a battery, but rather used selectively according to the information available on the chemical. These tests have been described in greater detail elsewhere [8, 12].

*1. Skin Tumor Induction in Mice.* The carcinogenicity of a limited number of chemicals and crude products can be revealed readily upon continuous application to the skin of mice, producing papillomas or carcinomas, or upon subcutaneous injection, yielding sarcomas.

*2. Pulmonary Tumor Induction in Mice.* Lung tumors can be induced in specific, sensitive strains of mice, especially the A/Heston strain and related strains like A/J in 15–20 weeks.

*3. Breast Cancer Induction in Female Sprague-Dawley Rats.* Exposure to carcinogens produces cancer in the mammary gland of young female random-bred Wistar and, to a greater degree, in Sprague-Dawley rats in less than 9 months.

*4. Altered Foci Induction in Rodent Liver.* Research in a number of laboratories has established that during liver carcinogenesis several distinct hepatocellular lesions precede the development of carcinomas. The earliest of these, the altered focus is characterized by several phenotypic abnormalities that permit its reliable and objective identification at early stages. With genotoxic carcinogens, foci have been detected within 3 weeks of carcinogen exposure and in high numbers by 12–16 weeks of exposure. Therefore, the recommended approach is that of continuous exposure to the test chemical with sampling for foci at weekly intervals up to 12 weeks.

In addition, the development of foci results in the release of gamma-glutamyltranspeptidase into the blood and thus measurement of this enzyme in the plasma can be used as another parameter to monitor the development of a carcinogen-induced abnormal population.

*5. Assays for Promoters.* In addition to providing further evidence for genotoxicity, limited bioassays in vivo can also be used to test for promoting substances. A genotoxic carcinogen that is active at a specific target organ, such as skin, breast, colon, urinary bladder or liver, is applied in small initiating doses after which the test compound is administered.

*Decision Point 2*

Because of the limited nature of these bioassays, negative results do not prove a lack of potential carcinogenicity. Thus, for a compound which displayed some activity in short-term tests or one with significant human exposure, chronic bioassay would be indicated.

The presence of positive results in two or more of the rapid in vitro tests reliably indicating genotoxicity and, also, a definite positive result in the limited in vivo bioassays would make a product highly suspect as a potential human carcinogenic risk. This is especially true if these results were obtained with moderate dosages and more so, if there was evidence of a good dose response, particularly as regards the multiplicity of the lung or mammary gland tumors.

The demonstration of promoting activity in any of the modified assays in the absence of genotoxicity indicates a specific type of hazard of the chemical deserving investigation as an epigenetic phenomenon.

*Stage D: Chronic Bioassay*

The decision point approach reserves chronic bioassay as the last resort in confirming questionable results in the more limited testing. It would also be applied for compounds that are inactive in the preceding stages but where extensive human exposure is likely. Such tests would also develop data on possible carcinogenicity through epigenetic mechanisms. In the latter situation, multispecies and dose-response data are most important if the data are to be applied to risk assessment. The elimination of unnecessary chronic testing for many chemicals by application of the decision point approach makes more extensive testing of suspected epigenetic agents economically feasible.

*Stage E: Final Evaluation*

At this stage all data generated must be included in the evaluation. If the sequence has led to chronic bioassay, definitely positive or negative data on carcinogenicity should have been obtained [7]. Nonetheless, the results of the in vitro short-term tests must be taken into account for an assessment of mechanisms of action and risk extrapolation to humans. Thus, convincing positive results in the in vitro tests together with documented in vivo carcinogenicity permits classification of the chemical as a genotoxic carcinogen. Such a chemical would have properties typical of these carcinogens, namely, the ability under some conditions to be effective as a single dose, cumulative effects, and synergism or at least additive effects with other genotoxic carcinogens. On the other hand, in the absence of convincing evidence for the genotoxicity of a chemical, but nonetheless indication of carcinogenicity in certain animal bioassays, it is possible that the chemical is an epigenetic carcinogen. The reliability of this conclusion depends upon the relevance of the in vitro tests.

In the case of ambiguous results in the chronic bioassay, clear positive results in the battery of short-term tests permit a judgment that the chemical is carcinogenic.

*Conclusions*

The decision point approach is designed to yield a stepwise progression of data acquisition that can be utilized for the systematic evaluation of the potential carcinogenic and mutagenic hazards of chemicals.

By following a logical sequence of tests, decisions can be made at key points in the series and preliminary and sometimes definite information can be obtained early as to whether or not a given material might or might not represent a carcinogenic or genetic risk. Moreover, the decision point approach can be readily integrated with other elements in toxicity testing.

The decision point approach not only has a sound science base but possesses economic advantages, an undeniable asset to government and industry, who share the expenses of bioassays, to ensure the safety of products on the market or to detect naturally occurring carcinogens. The same or better extent of public protection against the hazards of genotoxic agents can be secured by the systematic use of the decision point approach without conducting a chronic bioassay requiring several years and costing several hundreds of thousands of dollars. Also, because of the mechanistic understanding offered by the recommended tests, with findings available for research to extend the frontiers of knowledge, rather than the mere performance of large routine bioassays, greater benefits would accrue in the long run in the field of carcinogen screening where the basic methods used have not changed much in the past 20 years. A further benefit of the decision point approach is that nongenotoxic chemicals ultimately selected for bioassay based upon concern for human exposure can be tested over a more extensive dose range to delineate dose-response characteristics and possibly identify thresholds. Thus, the decision point approach provides an effective means of acquiring the essential information for safety evaluation of chemicals.

## *Acknowledgements*

This work was supported by grant CA-17613 from the National Cancer Institute. The authors thank Mrs. *Linda Stempel* and Ms. *Judy Benvin* for editorial assistance. This publication is dedicated to the founder of the American Health Foundation, Dr. *Ernst L. Wynder,* on the occasion of the 10th anniversary of the Naylor Dana Institute for Disease Prevention.

## *References*

Because of page limitations, literature citations have been restricted to primarily review articles and multiauthor monographs.

1 Grover, P.L.: Chemical carcinogens and DNA, vol. I, II; (CRC Press, Boca Raton 1979).

2 Hollstein, M.; McCann, J.; Angelosanto, F.; Nichols, W.: Short-term tests for carcinogens and mutagens. Mutat. Res. *65:* 133–226 (1979).

3 International Agency for Research on Cancer: Monograph on the evaluation of the carcinogenic risk of chemical to humans. Suppl. 2: Long-term and short-term screening assays for carcinogens: a critical appraisal. (International Agency for Research on Cancer, Lyon 1980).

4 Miller, J.; Miller, E.: in Emmelot, Kriek, Environmental carcinogenesis, pp. 25–50 (Elsevier, Amsterdam 1979).

5 Mishra, N.; Dunkel, V.; Mehlman, M.: Advances in modern environmental toxicology. vol. 1: Mammalian cell transformation by chemical carcinogens (Senate Press, New Jersey 1981).

6 Stich, H.F.; San, R.H.C.: Short-term tests for chemical carcinogens (Springer, New York 1981).

7 Tomatis, L.: The predictive value of rodent carcinogenicity tests in the evaluation of human risks. Annu. Rev. Pharmacol. Toxicol. *19:* 511–530 (1979).

8 Weisburger, J.H.; Williams, G.M.: Decision point approach to carcinogen testing; in Asher, Zervos, Symposium on structural correlates of carcinogenesis and mutagenesis, pp. 45–52 (Office of Science, Rockville 1978).

9 Weisburger, J.H.; Williams, G.M.: Carcinogen testing. Current problems and new approaches. Science *214:* 401–407 (1981).

10 Weisburger, J.H.; Williams, G.M.: Metabolism of chemical carcinogens; in Becker, Cancer; 2nd ed., vol. 1, pp. 241–333 (Plenum Press, New York 1982).

11 Weisburger, J.H.; Williams, G.M.: Chemical carcinogenesis; in Doull, Klaassen, Amdur, Toxicology: the basic science of poisons, pp. 84–138 (Macmillan, New York 1980).

12 Williams, G.M.; Weisburger, J.H.: Systematic carcinogen testing through the decision point approach. Annu. Rev. Pharmacol. Toxicol. *21:* 393–416 (1981).

13 Williams, G.M.; Kroes, R.; Waaijers, H.W.; Van de Poll, K.W.: The predictive value of short-term screening tests in carcinogenicity evaluation; in Applied methods in oncology, vol. 3 (Elsevier/North-Holland Biomedical Press, Amsterdam 1980).


Dr. G.M. Williams, Naylor Dana Institute for Disease Prevention,
American Health Foundation, Valhalla, NY 10595 (USA)

Safety Evaluation and Regulation of Chemicals. 1st Int. Conf., Boston, Mass., 1982, pp. 210–216 (Karger, Basel 1983)

# Background Variations of Tumor Incidence in Rodent Populations

*J. M. Ward*

Tumor Pathology and Pathogenesis Section, Laboratory of Comparative Carcinogenesis, Division of Cancer Cause and Prevention, National Cancer Institute, National Institutes of Health, Frederick, Md., USA

Differences in naturally-occurring tumor rates in a given strain of experimental rats or mice can readily be demonstrated between control groups in various experiments. The causes of these variations may be either simple or complex. Knowledge of the causes and characteristics of these variations would aid in evaluation of the bioassay and in drawing conclusions from them. It is the purpose of this paper to review some of the possible causes of variation of tumor rates.

## *Known Causes of the Variation of Tumor Rates*

The most obvious causes of variation in tumor rates include differences between species, strains, stock, sexes and diets (table I). Examples of these factors have been reviewed [14]. The effects of age, sex, genetics and statistical variation [2, 8] are well known. Less well studied have been the effects of diet [7], drinking water, environment, litter effects and cage position [4]. For example, the incidence of lymphoma in an experiment was associated with the level at which cages were placed on the rack [4].

## *The Pathology Examination as a Cause of Tumor Incidence Variation*

The extensiveness and quality of the pathology examination in an experiment can greatly affect the apparent incidence of tumors and even the outcome of a bioassay. The necropsy protocol, trimming of tissues, number and size of pieces of tissues taken, histopathology review and diagnosis affect the recorded tumor incidence. These aspects of tumor rate variation have been discussed in detail [14, 15]. More recently, a bioassay of Arcolor 1254 (PCB) was reevaluated in light of a limited pathology protocol performed originally [5]. Initially, only grossly visible stomach lesions were prepared for histologic review. The conclusion was made that carcinomas of the gastrointestinal tract may be associated with the administration of Arcolor 1254. After examining sections of almost all stomachs, a new conclusion noted a dose-related increase in the incidence of focal lesions in the glandular stomach, some of which were neoplastic and others of which were potentially preneoplastic. Thus, the conclusion varied according to the pathology protocol. On the other hand, gross findings in 22 bioassays provided adequate evidence of carcinogenicity when compared with microscopic findings, although the incidence of almost all tumors in control and treated rats was higher when tissues were examined microscopically [15]. The method and accuracy of tumor diagnoses may affect the tumor incidence. In a bioassay of dimethyl terephthalate, a re-evaluation of lung slides by a National Toxicology Program pathologist resulted in a doubling of the lung tumor incidence in control mice and an increased incidence of lung carcinomas in mice exposed to the chemical [6]. While the tumor incidence in the original NCI report was based on consensus diagnoses by several pathologists, the reevaluation was based on those of only 1 pathologist.

### *Variation in Tumor Incidence: Multiple Tumors or Tumors Associated with Tumors*

Tumors with interrelated etiologies or mechanisms may be associated. *Suzuki* et al. [10] showed that certain endocrine tumors tended to occur in rats with other endocrine tumors, a situation similar to the human endocrine adenomatoses. Thyroid C cell tumors in Sprague-Dawley rats were significantly associated with adrenal pheochromocytomas. Associated tumors may also include prolactinomas (pituitary chromophobe adenomas producing prolactin) and mammary tumors, especially fibroadenomas. Thus, if a control group has a high incidence of prolactinomas, one may expect a higher incidence of mammary tumors. Mononuclear cell leukemia in the F344 rat has recently been described as a large granular lymphocyte (LGL) leukemia with natural killer cell (NK) (cytotoxic) activity against other tumor cells in vitro [9]. Most F344 rats have 2–3 tumors per rat from 2 to 3 years of age. If the LGL leukemia had significant NK activity against other natural tumors

*Table I.* Potential causes of variation in tumor incidence of a strain or stock [discussed in ref. 14]

| | |
|---|---|
| 1 Age | 6 Statistical variation (binomial and nonbinomial) |
| 2 Sex | 7 Statistical analysis |
| 3 Animal husbandry | a Combination of tumor types |
| a Diet and water | b Statistical methods |
| b Cage density | c Use of concurrent, laboratory and historical controls |
| c Location on cage rack | 8 Unknown factors |
| d Animal room | a Genetic drift |
| e Survival and life span | b Year |
| f Diseases | c Room and laboratory (facility) differences (environment) |
| 4 Necropsy protocol | |
| 5 Method of histopathology diagnoses | |

*Table II.* Incidence of tumors in F344 rats with and without natural LGL leukemia[1]

| Tumor/organ | Males | | Females | |
|---|---|---|---|---|
| | with leukemia (73 rats) | without leukemia (167 rats) | with leukemia (64 rats) | without leukemia (259 rats) |
| Testes | 67 (91.7) | 161 (96.4) | N/A | N/A |
| Pituitary | 7 (9.5) | 22 (13.1) | 22 (34.3) | 96 (37) |
| Adrenal pheochromocytoma | 13 (17.8) | 25 (14.9) | 2 (3.1) | 9 (3.4) |
| Adrenal cortex | 2 (2.7) | 1 (0.5) | 2 (3.1) | 9 (3.4) |
| Thyroid C cell | 3 (4.1) | 10 (5.9) | 2 (3.1) | 11 (4.2) |
| Thyroid F cell | 2 (2.7) | 3 (1.7) | 1 (1.5) | 1 (0.4) |
| Mesothelioma | 1 (1.3) | 9 (5.3) | 0 | 1 (0.4) |
| Islets | 4 (5.4) | 6 (3.5) | 0 | 3 (1.1) |
| Liver | 8 (10.9) | 6 (3.5) | 3 (4.6) | 12 (4.6) |
| Mammary fibroadenoma | 1 (1.3) | 4 (2.3) | 9 (14.0) | 55 (21.2) |
| Mammary adenoma | 0 | 0 | 1 (1.5) | 7 (2.7) |
| Mammary carcinoma | 0 | 1 (0.5) | 0 | 5 (1.9) |
| Skin | 2 (2.7) | 8 (4.7) | 0 | 9 (3.4) |
| Fibroma (skin) | 7 (9.5) | 9 (5.3) | 1 (1.5) | 2 (0.8) |
| Preputial/clitoral gland | 3 (4.1) | 4 (2.3) | 0 | 8 (3.0) |
| Uterine stromal polyp | N/A | N/A | 9 (14.0) | 60 (23.1) |
| Other tumors | 2 (2.6) | 11 (6.1) | 1 (1.5) | 12 (4.6) |

N/A = Not applicable. Data in parentheses denote percentage of rats with tumors.

[1] Data from NCI-NTP Pathology Archival Repository and Carcinogenesis Bioassay Data System.

*Table III.* Variation in mean tumor incidence and range of tumor incidences in toxicology laboratories[1]

| Species | Sex | Tumor type or tissue | Mean tumor incidence for all laboratories | Range of mean tumor incidences in all laboratories[2] | Range of tumor incidences in control groups in all laboratories |
|---|---|---|---|---|---|
| Rat | M | pituitary gland | 20.3 | 7.5–31.2 | 0–65 |
| (F344) | F | pituitary gland | 40.0 | 33.6–58.6 | 5–80 |
| | M | leukemia | 16.9 | 9.1–23.6 | 0–54 |
| | F | leukemia | 11.0 | 7.5–15.4 | 0–32 |
| | M | adrenal pheochromocytoma | 10.7 | 4.9–16.0 | 0–30 |
| | F | adrenal pheochromocytoma | 3.2 | 1.1– 5.6 | 0–18 |
| | F | endometrial stromal polyp | 14.9 | 9.9–25.3 | 0–38 |
| | F | mammary fibroadenoma | 15.5 | 6.4–24.2 | 0–38 |
| Mouse | M | liver | 32.1 | 25.0–40.1 | 7–58 |
| (B6C3F1) | F | liver | 6.2 | 4.6– 9.7 | 0–21 |
| | M | lung | 17.0 | 10.6–21.9 | 0–45 |
| | F | lung | 6.0 | 3.6– 7.1 | 0–20 |
| | M | lymphoma/leukemia | 11.2 | 7.2–12.2 | 0–33 |
| | F | lymphoma/leukemia | 24.4 | 17.0–30.4 | 5–45 |

[1] From *Tarone* et al. [11]. There were 5 NCI Contract laboratories for mice and 6 for rats with 7–22 control groups in each laboratory. Experiments were usually terminated at 110–114 weeks of age.
[2] 72 rat and 54 mouse control groups, 20–55 animals per group.

in aging F344 rats, the incidence of these tumors might be reduced in rats with LGL leukemia. In a preliminary study, table II records the tumors found in rats with and without LGL leukemia in five 2-year bioassays. A few tumor types (mammary fibroadenoma, uterine stromal polyps and rare tumors) were found in lower incidence in rats with LGL leukemia. Since LGL leukemia was a major cause of death of rats prior to the termination of a typical 2-year bioassay [14], the decreased incidence of these tumors may be a result of a shorter survival time for these rats. A detailed evaluation of these findings will be reported.

*Table IV.* NCI bioassays with suggestive evidence of carcinogenicity[1]

| Species | Potential target site | Chemical | Sex | Incidence in high or low dose group, % | Program Historical Control Groups Incidence mean, %[2,3] | variation, %[3,4] |
|---|---|---|---|---|---|---|
| Rat | leukemia | acetohexamide | M | 29 | 16.9 | 0–54 |
| (F344) | preputial gland | *p*-anisidine HC1 | M | 15 | 1.4 | N/A[5] |
| | brain | 1-H-benzotriazole | M | 7 | 0.8 | 0–4[4] |
| | intestine | proflavine | M | 10 | 0.4 | N/A |
| | spleen | *p*-chloroaniline | M | 20 | 0.1 | N/A |
| | thyroid follicles | N, N[1]-dicyclohexylthiourea | M | 14 | 1.7 | N/A |
| | | Di-(2-ethylhexyl)-adipate | M | 55 | 32.1 | 15–58 |
| Mouse | liver | DCDD | M | 40 | 32.1 | 15–58 |
| (B6C3F1) | | dieldrin | M | 40 | 32.1 | 15–58 |
| | | proflavine | F | 62 | 6.2 | 0–21 |
| | | perthane | F | 23 | 6.2 | 0–21 |
| | | 2,5-dithiobiurea | F | 23 | 6.2 | 0–21 |
| | lung | styrene | M | 21 | 17.0 | 0–45 |
| | | dimethyl terephthalate | F | 35 | 6.0 | 0–20 |
| | | BHT | F | 20 | 6.0 | 0–20 |
| | | 1-H-benzotriazole | M | 27 | 17.0 | 0–45 |

1 NCI or NTP Carcinogenesis Bioassay reports are available from the National Technical Information Service, Springfield, Va. Also see *Chu* et al. [1].
2 From *Goodman* et al. [3] and *Ward* et al. [13].
3 From *Tarone* et al. [11].
4 From *Ward and Rice* [16].
5 Not Available.

### *Binomial and Nonbinomial Variations: The Correlated Binomial Heterogeneity Test*

A correlated binomial heterogeneity test [12] was used to evaluate the control tumor rates of F344 rats and B6C3F1 mice used in carcinogenesis bioassays in toxicology laboratories under contract to the National Cancer Institute [11]. The range in tumor incidence was striking but generally binomial (homogeneous) for most tumor types studied at a given laboratory (table III). Interlaboratory heterogeneity for most tumor types and intra-laboratory heterogeneity for rat pituitary tumors and endometrial stromal polyps were observed. The variation in control tumor rates at a given laboratory can be utilized as an aid in evaluation of bioassay results. Suggested methods for using this information are noted in *Tarone* et al. [11] and in the following section.

### *Application of Known Tumor Incidence Variation to Evaluation of Rodent Carcinogenesis Bioassays*

*Tarone* et al. [11] gave a few examples of the use of historical tumor incidence data to evaluate carcinogen bioassays. Based on the experience of scientists in the NCI and NTP programs, historical data are often used as a guide [1, 2]. Because of the limitations of rodent bioassays involving 50 animals per group, and the known tumor incidence variations in control rodent populations, the conclusions of some experiments will invariably be equivocal. Thus, in a review of 200 NCI chemical bioassays, the findings of the bioassays revealed that 20 chemicals were carcinogenic in both sexes of both species, 23 were carcinogenic in one sex of both species, 50 were carcinogenic in one species, 67 were not shown to be carcinogenic, 15 were inadequate for evaluation and 25 (12.5%) were suggestive of a carcinogenic effect [1]. Some of the latter 25 bioassays were suggestive because of unusually low concurrent control tumor incidences, effects which were not dose related or few 'induced' tumors (statistically not significant) in a site where spontaneous tumors are rare. Examples are given in table IV. Additional information would be needed to evaluate the carcinogenicity of these chemicals. While table IV shows variation in control groups of any NCI laboratory, the best comparison, not available for this table, would be that of the same laboratory in which the bioassay was performed [11].

### *References*


1 Chu, K.C.; Cueto, C., Jr.; Ward, J.M.: Factors in the evaluation of 200 National Cancer Institute Carcinogen Bioassays. J. envir. Path. Toxicol. *8:* 251–280 (1981).

2 Gart, J.J.; Chu, K.C.; Tarone, R.E.: Statistical issues in the interpretation of chronic bioassay tests for carcinogenicity. J. natn. Cancer Inst. *62:* 957–974 (1979).

3 Goodman, D.G.; Ward, J.M.; Squire, R.A.; Chu, K.C.; Linhart, M.S.: Neoplastic and nonneoplastic lesions in aging F344 rats. Toxicol. appl. Pharmacol. *48:* 237–248 (1978).

4 Lagakos, S.; Mosteller, F.: A case study of statistics in the regulatory process. The FD&C red 40 experiments. J. natn. Cancer Inst. *66:* 197–212 (1981).

5 Morgan, R.W.; Ward, J.M.; Hartman, P.E.: Aroclor 1254-induced intestinal metaplasia

and adenocarcinoma in the glandular stomach of F344 rats. Cancer Res. *41:* 5052–5059 (1981).

6 National Toxicology Program: Reevaluation by the National Toxicology Program of technical report NCI-CG-TR-121 entitled bioassay of dimethylterephthalate for possible carcinogenicity. Fed. Reg. *46:* 60654–60657 (1981).

7 Newberne, P.M.; McConnell, R.G.: Dietary nutrients and contaminants in laboratory animal experimentation. J. envir. Path. Toxicol. *4:* 105–122 (1980).

8 Peto, R.; Pike, M.C.; Day, N.E.; Gray, R.G.; Lee, P.N.; Parish, S.; Peto, J.; Richards, S.; Wahrendorf, J.: Guidelines for simple, sensitive significance tests for carcinogenic effects in long-term animal experiments; in Long-term and short-term screening assays for carcinogens. A critical appraisal. IARC suppl. 2, pp. 311–426 (IARC, Lyon 1980).

9 Reynolds, C.W.; Ward, J.M.; Denn, A.C., III; Bere, E.W., Jr.: Identification and characterization of large granular lymphocyte (LGL) leukemias in F344 rats; in Herberman, Natural immunity, vol. 2 (Academic Press, New York, in press, 1982).

10 Suzuki, H.; Molt, H.; Kimmerle, G.: Spontaneous endocrine tumors in Sprague-Dawley rats. J. Cancer Res. clin. Oncol. *95:* 187–196 (1979).

11 Tarone, R.E.; Chu, K.C.; Ward, J.M.: Variability in the rates of some common naturally occurring tumors in Fischer 344 rats and (C57BL/6N × C3H/HeN)F1 (B6C3F1) mice. J. natn. Cancer Inst. *66:* 1175–1181 (1981).

12 Tarone, R.E.: Testing the goodness of fit of the binomial distribution. Biometrika *66:* 585–590 (1979).

13 Ward, J.M.; Goodman, D.G.; Squire, R.A.; Chu, K.C.; Linhart, M.S.: Neoplastic and nonneoplastic lesions in aging (C57BL/6N × C3H/HeN)F1 (B6C3F1) mice. J. natn. Cancer Inst. *63:* 849–854 (1979).

14 Ward, J.M.: Background data and variations in tumor rates of control rats and mice; in Homburger, Prog. exp. Tumor Res., vol. 26, pp. 241–258 (Karger, Basel 1983).

15 Ward, J.M.; Reznik, G.: Refinements of rodent pathology and the pathologist's contribution to evaluation of carcinogenesis bioassays; in Homburger, Prog. exp. Tumor Res., vol. 26, pp. 266–291 (Karger, Basel 1983).

16 Ward, J.M.; Rice, J.M.: Naturally occurring and chemically induced brain tumors of rats and mice in carcinogenesis bioassays. Ann. N.Y. Acad. Sci. (in press, 1982).

Dr. J.M. Ward, Laboratory of Comparative Carcinogenesis, National Cancer Institute, Frederick, MD 21701 (USA)

Safety Evaluation and Regulation of Chemicals. 1st Int. Conf., Boston, Mass., 1982, pp. 217–224 (Karger, Basel 1983)

# Early Detection of Lesions in Experimental Animals

*Hildegard M. Reznik-Schüller*

Chemical Carcinogenesis Program, NCI-Frederick Cancer Research Facility, Frederick, Md., USA

A large number of chemicals with toxic, mutagenic and carcinogenic effects on the mammalian organism have been identified in recent years. The vast majority of such compounds are linked with modern human life-style and behavior. To keep potential hazards to man as low as possible, it is vital to detect the harmful effects of chemicals as early as possible. A number of in vitro and in vivo systems have been developed for this purpose. While most of the in vitro studies are short-term tests which yield results within 2 weeks, the conventional in vivo investigations in experimental animals may last for several years until conclusive data become available. However, chemically-induced lesions may be detected much earlier in the animal models if a methodology more sophisticated than routine histopathology is applied. In this article, such possibilities for the early detection of lesions in experimental animals will be exemplified.

## *Methodology*

### *Quantitative Morphometric Analysis on Tissue Sections for Light and Electron Microscopy*

Chemicals which are toxic, mutagenic or carcinogenic exert such effects by interacting with the cells of the host organism. At an early stage, before pathological damage occurs, such interactions are reflected by quantitative changes in the affected cells and tissues. At the light microscopic level, such changes include alterations in the sizes and shapes of cells and nuclei, variations in the nucleus/cytoplasm ratio, alterations in the distribution pattern of cell types, and frequency of early lesions (e.g. cytoplasmic swelling, altered staining characteristics, patterns of histochemical reactions). The induced cellular changes are conveniently recorded and processed for statistical analysis by the aid of modern image analysis systems. In principle, such instruments consist of a tablet on which the light microscopic image may be projected directly or on which photomicrographs are placed. With an electronic pen the structures of interest (e.g. cells, nuclei) are then traced (for analysis of size and shape) or counted (e.g. for analysis of frequency distributions). The built-in microprocessor and attached computers then convert the entered data into area in $\mu m^2$ and frequency distributions of counted events (in %), store the data and compare data among groups by selected statistical tests [more detailed description in ref. 4].

Further insight into functional changes of the affected cell types may be obtained by morphometric analysis of their ultrastructure. Electron micrographs taken at a standard magnification are mounted on the tablet of the image analyzer. Changes in the ultrastructure of cell organelles as well as increase or decrease in their volume per cell can thus be recorded and evaluated.

*Autoradiography*

The distribution and site of binding of the tested chemical in the animal can be monitored by autoradiographic techniques using the radioactively labeled compound. Wholebody autoradiography done at various time intervals after administration of the labeled chemical to the animal, reveals the distribution of the chemical in the body as a function of time. It also shows if there are any organs in which the chemical accumulates and/or persists for a longer time. Light microscopic autoradiography will subsequently elucidate if binding of the chemical and/or its metabolites occurs in any particular cell types. Ultimately, electron-microscopic autoradiography may then be applied to examine if binding occurs at any preferred intracellular sites (e.g. endoplasmic reticulum, mitochondria, heterochromatin). The latter technique, however, does not yield results as quickly (exposure times are up to 6 months) as the other procedures mentioned in this article and, hence, are not recommended if quick data are required.

## *Examples for Early Detection of Lesions in Different Organ Systems*

*Respiratory Tract*

Early chemically-induced lesions in the respiratory tract may be found at the epithelial surfaces (e.g. cilia) or in intracellular components. For the rapid quantitative detection of changes at the cell surfaces, scanning electron

*Fig. 1.* Scanning electron micrograph demonstrating middle portion of Syrian golden hamster trachea after intratracheal instillation of automobile exhaust condensate: large portions of the epithelial surface are denuded of cilia. × 2,450. (The author gratefully acknowledges the aid of Philips, Eindhoven, The Netherlands, in whose laboratory the picture was made.)

microscopy is most useful. The most common lesions, induced by inhaled noxious agents, are alterations of the respiratory cilia in the air conducting ways. Such changes, which have also been reported after treatment with intratracheally and systemically administered carcinogens [2] may be a reduction in the number of cilia per unit surface area (fig. 1) or deformation of the cilia (fig. 2). Such alterations, which are also commonly found in heavy smokers (fig. 3) and human patients with bronchial carcinoma, result in an impairment of the important mucociliary defense mechanisms and, hence, may pave the way for more serious damage from subsequently inhaled substances.

As to the intracellular changes, these are best to be detected (after 1–2 weeks of treatment) by transmission electron microscopy including morphometry while changes detectable by light microscopy usually require longer treatment periods (e.g. 4–6 weeks) to become apparent. Such changes appear to be especially prominent after treatment with chemicals which require metabolic activation in the host organism. Apparently, the target cell types respond to the elevated need for metabolically capable organelles by

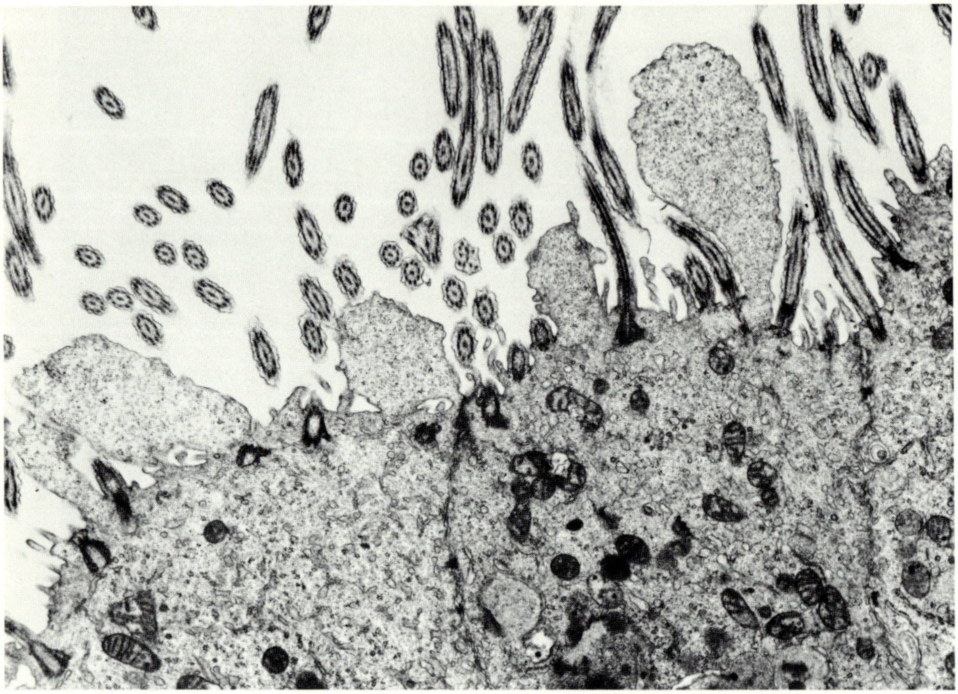

*Fig. 2.* Transmission electron micrograph of respiratory epithelium in Syrian golden hamster trachea after treatment with diethylnitrosamine: deformation of cilia is common. ×10,500.

*Table I.* Morphometric analysis of tracheal mucous cells and bronchial Clara cells of Syrian golden hamsters treated for 2 weeks with diethylnitrosamine

| Measured organelle (% volume/cell) | Mucous cells | | Clara cells | |
|---|---|---|---|---|
| | controls | dosed | controls | dosed |
| Rough endoplasmic reticulum | 25.3 ± 3.8% | 38.7 ± 2.8% | 5.4 ± 2% | 2.0 ± 1% |
| Smooth endoplasmic reticulum | 0 | 0 | 35.3 ± 4.3% | 45.2 ± 3.4% |
| Mitochondria | 12.1 ± 2.2% | 14.5 ± 3.1% | 8.2 ± 3.1% | 10.3 ± 2.8% |
| Secretion granules | 14.8 ± 3.1% | 8.1 ± 2.5% | 2.4 ± 1.2% | 0.5 ± 1% |
| Lysosomes | 1.5 ± 1% | 3.2 ± 0.8% | 1.8 ± 1.1% | 0.3 ± 0.7% |
| Nucleus | 28 ± 10% | 31 ± 7.2% | 30 ± 5.8% | 33 ± 6.8% |

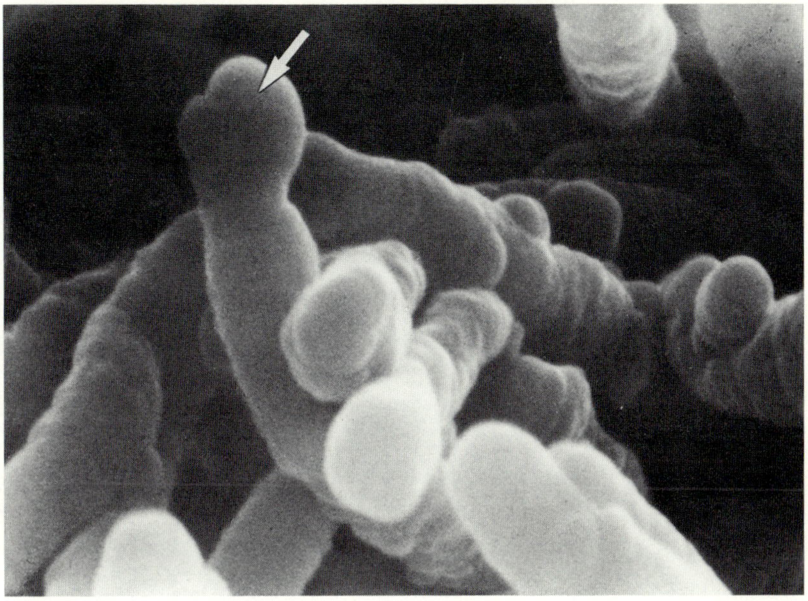

*Fig. 3.* Scanning electron micrograph from bronchial biopsy sample in human smoker: balloon-like deformation of ciliary tip (arrowed) is common. ×39,000. (The author gratefully acknowledges the aid of Philips, Eindhoven, The Netherlands, in whose laboratory the picture was made.)

increasing their number and/or size. A number of nitrosamines which are lung carcinogens have been found to induce proliferation of endoplasmic reticulum (table I) in bronchial mucous and Clara cells in hamsters [4, 6] and rats [3]. Autoradiographic studies with the radioactively-labeled compounds revealed the ER in these cells as the major site of binding in the lungs.

*Liver*

The liver is one of the principal organs responsible for detoxification and metabolization reactions. Naturally, this organ is often affected in some way by chemicals administered to animals. The smooth endoplasmic reticulum of the hepatocytes has long been recognized as one of the most important intracellular sites for detoxification of chemicals. As in the respiratory tract, this organelle may increase or decrease in accordance with the needs of the cell. A typical example for this phenomenon is the well-known proliferation of smooth endoplasmic reticulum after treatment with phenobarbital. In

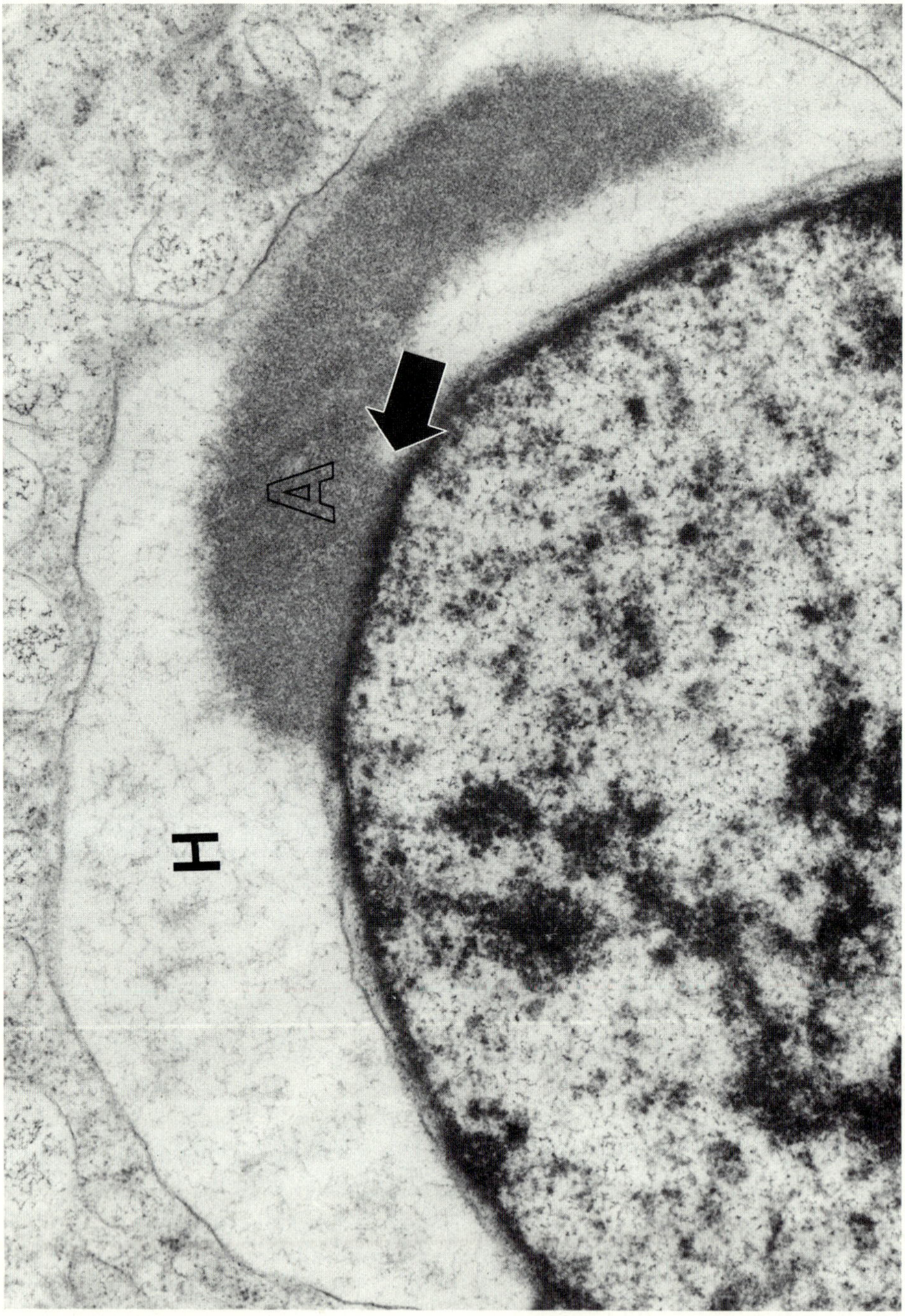
A
H

*Table II.* Morphometric analysis of portal hepatocyte in F344 rats treated for 2 weeks with methapyrilene hydrochloride

| Organelles | Controls (% volume/cell) | Dosed (% volume/cell) |
|---|---|---|
| Rough endoplasmic reticulum | 24.1 ± 3.1 | 11.2 ± 2.5 |
| Smooth endoplasmic reticulum | 45.3 ± 4.8 | 17.5 ± 3.1 |
| Mitochondria | 8.6 ± 3.4 | 29.7 ± 2.8 |
| Lipid | 0.3 ± 0.2 | 0.2 ± 0.1 |
| Lysosomes | 1.4 ± 0.8 | 1.8 ± 0.7 |
| Nucleus | 13.1 ± 0.5 | 16.4 ± 1.2 |

addition, a number of non-genotoxic agents have been reported to induce proliferation of peroxisomes in hepatocytes. In my own laboratory, we recently identified methapyrilene as a mitochondrial proliferator in hepatocytes [5]. Methapyrilene is also considered a non-genotoxic agent and is a powerful liver carcinogen in the rat [1]. After administration of 1,000 ppm in the food for only 2 weeks, methapyrilene induced a pronounced proliferation of mitochondria in periportal hepatocytes (table II). These cells were identified as the origin of hepatocellular carcinomas in a serial sacrifice study. Moreover, autoradiography with [$^3$H]-methapyrilene revealed mitochondria as the principal site of binding in hepatocytes.

*Testes*

As is generally accepted, the male and female reproductive organs are fairly sensitive to noxious agents. Especially the complicated process of spermiogenesis, which consists of 17 defined stages, is easily affected. However, the subtle changes are usually not detected by light microscopy although they may ultimately lead to a severe decrease in fertility. The head cap and acrosome of the spermatids from which the head of the mature sperm develops, appears to be highly sensitive to the influence of noxious agents. In a number of experiments conducted in my laboratory, these organelles were found damaged after a short period of systemic treatment with chemical carcinogens as well as after administration of compounds with a generally low toxicity. Two synthetic retinoids, for example, which did not cause patholog-

*Fig. 4.* Transmission electron micrograph of headcap in spermatid of Syrian golden hamster treated with a synthetic retinoid: the acrosome (A) is deformed. × 44,500.

ical changes in any other organs of Syrian golden hamsters, were found to induce atrophy of the seminiferous tubules and/or alterations of the acrosomes and head caps [7]. These changes were deformations of head cap and acrosomes (fig. 4) as well as swelling and/or disruption of their membranes. As evident from figure 4, such changes are easily detected by electron microscopy while they cannot be found under the light microscope.

## *Comments*

This brief summary of possibilities for the early detection of lesions in experimental animals demonstrates that such in vivo systems can yield data on adverse effects of chemicals within a relatively short period of time. With respect to their duration, such studies are quite comparable with short-term in vitro tests. The costs, too, are not very high once the equipment has been purchased. When applied in a meaningful combination, the outlined methods would seem to offer additional possibilities for the screening of chemicals for adverse effects in the mammalian organism.

## *References*

1 Lijinsky, W.; Reuber, M.B.; Blackwell, B.N.: Liver tumors induced in rats by oral administration of the antihistaminic methapyrilene hydrochloride. Science *209:* 817–819 (1980).
2 Reznik-Schüller, H.M.: Ciliary alterations in hamster respiratory epithelium after exposure to carcinogens and cigarette smoke. Cancer Lett. *1:* 7–13 (1975).
3 Reznik-Schüller, H.M.; Gregg, M.: Pathogenesis of lung tumors induced by N-nitrosoheptamethyleneimine in F344 rats. Virchows Arch. Abt. A Path. Anat. *393:* 333–345 (1981).
4 Reznik-Schüller, H.M.; Hague, B.F., Jr.: A morphometric study of the pulmonary Clara cell in normal and nitrosoheptamethyleneimine-treated European hamsters. Expl. Pathol. *18:* 366–371 (1980).
5 Reznik-Schüller, H.M.; Lijinsky, W.: Morphology of early changes in liver carcinogenesis induced by methapyrilene. Arch. Tox. *49:* 79–83 (1981).
6 Reznik-Schüller, H.M.; Reznik, G.: Experimental pulmonary carcinogenesis. Int. Rev. exp. Pathl. *20:* 211–281 (1979).
7 Stinson, S.F.; Reznik-Schüller, H.M.; Reznik, G.; Donahoe, R.: Atrophy induced in the tubules of the testes of Syrian hamsters by two retinoids. Toxicology *17:* 343–353 (1980).

H.M. Reznik-Schüller, DVM, PhD, Chemical Carcinogenesis Program, National Cancer Institute–Frederick Cancer Research Facility, PO Box B, Frederick, MD 21701 (USA)

Safety Evaluation and Regulation of Chemicals. 1st Int. Conf., Boston, Mass., 1982, pp. 225–232 (Karger, Basel 1983)

# Standardizing Syrian Hamsters for Toxicology

*F. Homburger, C.G. Van Dongen, R. Adams, E. Soto*
Bio-Research Institute, Cambridge, Mass., USA

There is a trend today toward short in vitro tests and a belief that these may eventually replace the traditional safety evaluation in intact animals. The validity of these conjectures is debatable on several grounds. From a practical point of view, it is likely that, if only for political reasons and because of the slowness and inertia of bureaucratic and legislative developments, toxicologists will for a considerable time to come be forced to employ biossays in vivo for safety evaluations, whether they like it or not.

It therefore behooves us to make every effort to improve, sharpen and shorten these time-honored procedures which we are conducting now much as we did when they were first formalized in 1959 in the much referred to *Appraisal of the Safety of Chemicals in Foods, Drugs and Cosmetics* published then by the Association of Food and Drug Officials of the United States [8].

Even the excellent conference on *Carcinogenesis Testing of Chemicals,* chaired in 1973 by our keynoter, *Golberg* [3], the *Guidelines for Carcinogen Bioassay in Small Rodents* of the NCI of 1975 [9], the *Good Laboratory Practices Regulations* of the FDA of December 1978 [4] and the report of the Interagency Regulatory Liaison Group, Work Group on Risk Assessment of 1979 [7], did not significantly alter the relatively crude method which we are using today, single-caging hundreds of rats and mice, measuring carefully many parameters and allowing the animals to live out most of their natural life span before we dissect them meticulously and search by light and electron microscopy sometimes, for the induced or absent tumors.

In the end we cannot agree in some cases, such as the liver lesions of mice, what does and what does not constitute cancer, and often we feel insecure when no new types of tumors arise but those normally occurring in controls are increased in numbers in the treated animals.

Improvements of our methods could come from a variety of approaches, some of which will be discussed in this conference.

Dr. *Williams* has already told us about the more rational selection of certain bioassays for specific purposes in safety evaluation.

Dr. *Ward's* data permit more precise evaluation of results in light of background tumor incidences. *Hildegard Reznik-Schüller* approaches the problem from the morphologist's point of view and *Mazué* will discuss the use of tumor markers, some of which were also introduced by *Solt* in yesterday's session. Another possible improvement is the use in bioassays for earlier tumor detection of electronic imaging of X-rays to reduce the dose, and of computer-assisted tomography (CAT scanning) for high resolution and clear anatomical detail in selected planes of the body.

The sensitivity of carcinogenesis bioassay could be increased by the introduction of other species into the spectrum of our in vivo test procedures. Ideally, these should be small animals (thus saving test compound), free of spontaneous cancers but known to be susceptible to carcinogens when these are administered and having a life-span comparable to that of the conventionally used rodents.

Syrian hamsters meet these requirements. Some who have attempted to use hamsters for lifetime carcinogenesis bioassay have expressed reservations, such as that the animals are hard to handle, subject to a strange disease called 'wet tail' or more scientifically, hyperplastic ileitis, and with a natural life-span which is shorter than that of rats or mice.

In a new first generation hybrid (15.16 sire vs. 87.20 dam), BIO F1D Alexander, all of these objections have been overcome and it is in our view an ideal animal for carcinogenesis bioassay. Data on the response of its parental strains to carcinogens administered by gavage and subcutaneously have been published and showed both strains to be susceptible to carcinogens [6]. One of them, the BIO 15.16 strain is also highly responsive to carcinogenesis in the larynx with inhaled cigarette smoke [1]. The other, BIO 87.20 does not develop this neoplasm when similarly exposed. Both parental strains have been highly inbred for many generations. Injection of carcinogens subcutaneously have shown the first generation hybrid BIO FlD Alexander to be as susceptible to polycyclic hydrocarbons as the parental strains [5]. Preliminary smoke inhalation studies in hybrids revealed considerable susceptibility to nicotine (to which the parental strains are resistant) [unpubl. data] and hence the hybrid is less suitable than its parents for tobacco smoke studies.

Table I shows the tumor incidence found in 100 CD1 Sprague-Dawley

*Table I.* Comparison of spontaneous tumor incidence in rats and Syrian hamsters and consequences for statistical evaluation of carcinogenesis studies

| | Number of animals | Total tumors, % | Percentages of various tumors[1] | | | | |
|---|---|---|---|---|---|---|---|
| | | | pituitary | mammary gland | bladder | lung | liver |
| Rats[2] | 111 | 59 | 32 (54) | 22 (41) | 5 (18) | 2 (13) | 2 (13) |
| Hamsters[3] | 100 | 4 | 0 (8) | 0 (8) | 0 (8) | 0 (8) | 1 (10) |

[1] Figures in parentheses indicate percentage of tumors required to show the significance of tests at the 1% level (National Bureau of Standards Handbook 91, August 1963).
[2] Rat data from Bio-Research Contract No. NIH-NCI-E68-1311 (lifetime).
[3] Hamster data from recent study of 100 male BIO 15.16 hamsters (lifetime).

*Table II.* Incidence (%) of animals with malignant tumors following MC feeding; histological diagnosis in 11 to 29 males and females of each strain

| Strain | Sex | Range of age at autopsy, weeks[1] | Number of survivors | % of autopsied animals with carcinomas of | | | | | | |
|---|---|---|---|---|---|---|---|---|---|---|
| | | | | fore stomach | stomach | small intestine | large intestine | ovary | uterus | mammary gland |
| RB | ♂ | 41–43 | 14 | 0 | 0 | 0 | 0 | | | |
| | ♀ | 41–47 | 21 | 14* | 0 | 25 | 25 | 50** | 24 | 52* |
| 4.22 | ♂ | 37–69 | 22 | 20* | 5 | 50* | 0 | | | |
| | ♀ | 35–48 | 18 | 25* | 5 | 45* | 10 | 33 | 39** | 61** |
| 15.16 | ♂ | 42–61 | 23 | 10 | 0 | 39* | 50* | | | |
| | ♀ | 42–72 | 29 | 4 | 3 | 12** | 0 | 25 | 3 | 86 |
| 54.7 | ♂ | 39–59 | 14 | 26* | 20* | 25 | 20 | | | |
| | ♀ | 37–53 | 24 | 12 | 15** | 33 | 12 | 38 | 42** | 91** |
| 82.73 | ♂ | 40–69 | 21 | 5 | 5 | 5 | 20 | | | |
| | ♀ | 40–50 | 22 | 20* | 5 | 20 | 13 | 20 | 32 | 82 |
| 86.93 | ♂ | 38–57 | 24 | 8 | 8 | 25 | 12 | | | |
| | ♀ | 37–48 | 25 | 4 | 3 | 0 | 4 | 56* | 12 | 80 |
| 87.20 | ♂ | 33–48 | 11 | 20* | 40* | 45* | 65* | | | |
| | ♀ | 34–43 | 23 | 5 | 12 | 12** | 33** | 21 | 4 | 87 |

[1] Age at autopsy = 13 weeks of age (beginning of feeding) + 17 weeks of feeding + survival thereafter.
Significant at * $p = 0.01$, ** $p = 0.05$.

*Table III.* Spontaneous tumor incidence in the F1D Alexander female: gross observations at autopsy in two separate groups

| Organ | Group 1 (n = 90) | Group 2 (n = 90) | Total with tumor | |
|---|---|---|---|---|
| | | | n | % |
| Kidney | | 3 | 3 | 1.7 |
| Adrenal | 18[1] | 10[2] | 28 | 15.6 |
| Thyroid | 1 | | 1 | 0.6 |
| Trachea | 1 | | 1 | 0.6 |
| Lung | | 2 | 2 | 1.1 |
| Pancreas[3] | | 3 | 3 | 1.7 |
| Liver | 2 | 4 | 6 | 3.3 |
| Small intestine | 1 | 3 | 4 | 2.2 |
| Large intestine[4] | 5 | 8 | 13 | 7.2 |
| Cecum[4] | 11 | 19 | 30 | 16.7 |
| Thymus | 1 | | 1 | 0.6 |
| Lymph nodes | 2 | | 2 | 1.1 |
| Spleen[5] | 3 | | 3 | 1.7 |
| Ovary | 3 | 1 | 4 | 2.2 |
| Oviduct | | 1 | 1 | 0.6 |
| Uterus and Cervix | 3 | 3 | 6 | 3.3 |
| Skin | 2[6] | 2 | 4 | 2.2 |
| Bone | | | | |
| Thoracic tumor | | 1 | 1 | 0.6 |
| Abdominal tumor | 1 | | 1 | 0.6 |
| Subcutaneous tumor | 2 | | 2 | 1.1 |

[1] Excludes 'cyst'; includes 'enlarged' and 'possible tumor'.
[2] Excludes 'cyst' and 'enlarged'.
[3] Red (pink) nodule(s) on surface.
[4] Nodules in the colon and the cecum frequently found together in the same hamsters.
[5] One 'enlarged spleen'; other instances described as 'white' or 'pale' nodules.
[6] One 'nodule on lower lip'.

rats (Charles River Breeding Laboratories) at the end of a lifetime bioassay and compares this with the tumors found in 100 BIO 15.16 inbred Syrian hamsters having lived nearly as long. There is a much lower incidence of spontaneous tumors in the hamsters. In parentheses are shown the incidences of similar tumors which would be required to demonstrate a significantly increased tumor incidence in test animals. Obviously, far fewer tumors need be induced in hamsters than in rats to demonstrate the carcinogenic potency of test substances. This is especially true of the types of neoplasms most frequently induced with administered carcinogens, namely mammary and gastrointestinal cancers, which are not seen spontaneously in these hamsters.

Table II illustrates the types of tumors which can be induced by the intragastric administration of methylcholanthrene in various inbred strains of hamsters [6]. Obviously, a low incidence of spontaneous tumors would be of no use for improvement of carcinogenesis bioassays if it were accompanied by resistance to administered carcinogens. There exists, fortunately, in the Syrian hamster, the paradox already noted by *Van Hoosier* et al. [10] of high susceptibility to administered carcinogen, yet low incidence of spontaneous tumors.

Table III reveals the autopsy findings in males and females of the BIO F1D Alexander hybrid. It must be noted that the females die much before the males, apparently from a combination of nephrosis and generalized amyloidosis, diseases which are much less frequent in the longer lived males. From gross necropsies it would appear that there are adrenal tumors and intestinal tumors. For histopathology see [6a].

Table IV reveals that histological examination does not confirm the adrenal enlargements noted at autopsy and classifies the intestinal lesions as cystic hyperplasias. The only neoplasms found in control animals are lymphomas, adrenal neoplasms and uterine tumors. It is noteworthy that those tumors frequently induced in hamsters by known carcinogens, namely mammary cancers and gastrointestinal tumors, are not seen spontaneously. There are also no pituitary, testicular or hepatic neoplasms.

The use of any new species must of course be preceded by pharmaco- or toxicokinetic studies which demonstrate that there are no basic differences between human and hamster metabolism of the test substance. When this is found to be the case, hamsters of well-defined genetic background and, specifically, the BIO F1D Alexander hybrid, may be the ideal animals for in vivo carcinogenesis bioassay.

*Chvedoff* et al. [2] of the French Pfizer Laboratories in Amboise, France, recently stated: 'The Syrian hamster may have advantages to offer over both rats and mice as a test system for evaluation of carcinogenic potential in the rodent life-span bioassay.' The acceptance of this species can greatly increase the sensitivity of these assays.

The performance of high resolution Mini-CAT scans at certain intervals, may, in the future, reduce the time needed to detect tumors in such tests.

*Table IV*. Histology of the normal F1D Alexander Syrian hamsters

| | | 87 males | | 87 females | |
|---|---|---|---|---|---|
| | | n | % | n | % |
| *Non neoplastic pathology* | | | | | |
| Kidney | tubular dilatation | 51 | 58.6 | 49 | 56.3 |
| Liver | fatty change | 5 | 5.7 | 7 | 8.0 |
| | cystic | 43 | 49.4 | 43 | 49.4 |
| Heart | thrombosis | 23 | 26.4 | 28 | 32.2 |
| Colon, cecum | hyperplasia, cystic | 29 | 33.3 | 10 | 11.5 |
| | polyp | 4 | 4.6 | | |
| Lung | acute inflammation | 7 | 8.0 | 8 | 9.2 |
| | chronic inflammation | 2 | 2.3 | 4 | 4.6 |
| | focal adenomatosis | 7 | 8.0 | 1 | 1.1 |
| Testes | atrophy | 49 | 56.3 | | |
| Thyroid | cyst | 2 | 2.3 | 5 | 5.7 |
| Ovary | cyst | | | 3 | 3.4 |
| Pancreas[1] | atrophy | 4 | 4.6 | | |
| Amyloidosis | | 5 | 5.7 | 45 | 51.7 |
| Arteritis | | | | 1 | 1.1 |
| *Malignant tumors* | | | | | |
| Lymphoma | | 9 | 10.3 | 5 | 5.9 |
| Adrenal cortical carcinoma | | 11 | 12.6 | 2 | 2.3 |
| s.c. sarcoma | | 2 | 2.3[2] | 1 | 1.1 |
| Colon or cecum carcinoma | | 2 | 2.3 | | |
| Uterine carcinoma[3] | | | | 4 | 4.6 |
| Pancreatic carcinoma | | | | 1 | 1.1 |
| Salivary gland carcinoma | | | | 1 | 1.1 |
| Melanoma | | | | 1 | 1.1 |
| *Benign tumors* | | | | | |
| Adrenal adenoma and hyperplasia | | 33 | 37.9 | 2 | 2.3 |
| Lymph node hemangioma | | 1 | 1.1 | | |
| Splenic hemangioma | | 1 | 1.1 | | |
| Pancreas (islet cells)[4] | | 5 | 5.7 | 2 | 2.3 |
| Pituitary adenoma | | | | 1 | 1.1 |
| Thecoma | | | | 2 | 2.3 |
| Leiomyoma | | | | 1 | 1.1 |
| Salivary gland adenoma | | | | 2 | 2.3 |
| Thyroid adenoma | | 2 | 2.3 | 1 | 1.1 |
| Histiocytoma | | 1 | 1.1 | | |

[1] Three instances of cyst.
[2] One sarcoma of the jaw; one of the peritoneum.
[3] One female had a primary uterine sarcoma as well.
[4] One additional female with pancreatic adenoma (not islet cell) and two with islet cell hyperplasia.

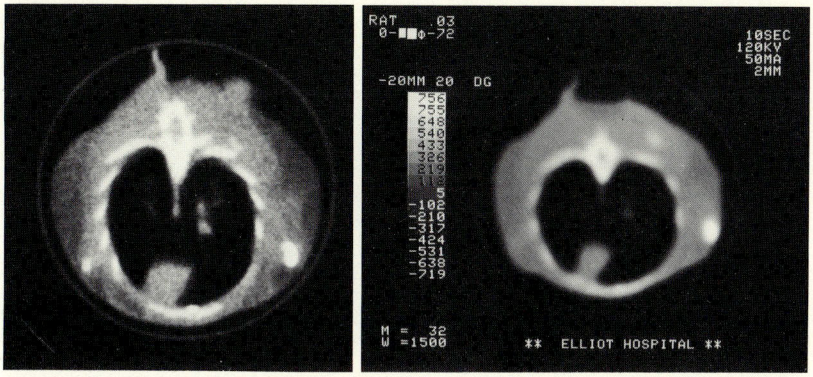


*Fig. 1.* *a* Mini-CAT scan of a Syrian hamster. *b* Unsatisfactory image obtained with clinical CAT scan instrument.

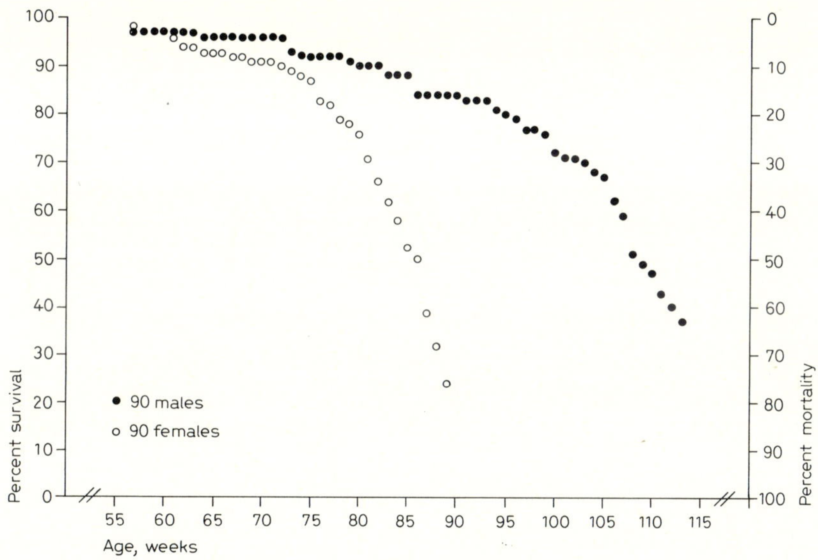


*Fig. 2.* Normal survival curves of BIO F1D Alexander animals.

Yesterday, at 5 AM in the morning, the physicists at American Science and Engineering under the leadership of Dr. *Roderick Swift* at Cambridge, Mass., recorded the first Mini-CAT scan of a Syrian hamster (fig. 1). This feat may well do for in vivo toxicology assays what Roentgen's discovery did for medicine. It is, however, possible that this development will again be outdated by the rapidly moving technology of nuclear magnetic resonance imaging.

Figure 2 shows the survival curves of males and females of BIO F1D Alexander animals. Females die earlier than males (see above). The males live longer than any of our inbred hamster lines.

## *References*


1 Bernfeld, P.; Homburger, F.; Russfield, A.B.: Strain differences in the response of inbred Syrian hamsters to cigarette smoke inhalation. J. natn. Cancer Inst. *53:* 1141–1157 (1974).

2 Chvedoff, M.; Dancla, J.L.; Faccini, J.M.; Greaves, P.; Perreaud, J.; Taradach, C.: The use of Syrian golden hamsters in chronic toxicity and carcinogenicity testing. Abstr. Ann. Meet. Brit. Soc. Toxicology, 1982.

3 Golberg, L.: Carcinogenesis testing of chemicals (CRC Press, Cleveland 1973).

4 Good Laboratory Practices Regulations: Federal Register, Dec. 22, 1978.

5 Homburger, F.; Adams, R.A.; Soto, E.; Van Dongen, C.G.: Susceptibility and resistance to chemical carcinogens in inbred Syrian hamsters. Prog. exp. Tumor Res., vol. 24, pp. 215–221 (Karger, Basel 1979).

6a Homburger, F.; Adams, R.A.; Bernfeld, P.; Van Dongen, C.G.; Soto, E.: A new first-generation hybrid Syrian hamster, BIO F1D Alexander for in vivo carcinogenesis bioassay as a third species or to replace the mouse. Surv. Synth. Path. Res. *1:* 127–134 (1983).

6b Homburger, F.; Hsueh, S.-S.; Kerr, C.S.; Russfield, A.: Inherited susceptibility of inbred strains of Syrian hamsters to induction of subcutaneous sarcomas and mammary and gastrointestinal carcinomas by subcutaneous and gastric administration of polynuclear hydrocarbons. Cancer Res. *32:* 360–366 (1972).

7 Interagency Regulatory Liaison Group: Testing Standards and Guidelines Group (US Consumer Products Safety Commission, US EPA, FDA, Food Safety and Quality Service of DA, PSHA): Scientific bases for identification of potential carcinogens and estimation of risk. J. natn. Cancer Inst. *63:* 241–268 (1979).

8 Lehman, A.J.: Appraisal of the safety of chemicals in foods, drugs and cosmetics: Association of Food and Drug Officials of the United States, Baltimore, Md., 1959 (Revision of an article from Food, Drug cosmet. Law J. *10:* 679, 1955).

9 Sontag, J.M.; Page, N.P; Saffiotti, U.: Guidelines for carcinogen bioassay in small rodents, (National Cancer Institute, NIH, Bethesda 1975).

10 Van Hoosier, G.L., Jr.; Spjut, H.J.; Trentin, J.J.: Spontaneous tumors of the Syrian hamster. Observations in a closed breeding colony and a review of the literature; in Defining the laboratory animal, (National Academy of Sciences, Washington 1971).


Dr. F. Homburger, Bio-Research Institute, Inc., 9 Commercial Avenue,
Cambridge, MA 02141 (USA)

Safety Evaluation and Regulation of Chemicals. 1st Int. Conf., Boston, Mass., 1982, pp. 233–242 (Karger, Basel 1983)

# Recent Developments in Inhalation Toxicology: Evaluation of Selected Short-Term Endpoints following 'Nose-Only' Exposure of Rodents[1]

*C.J. Henry, K.K. Hwang, K.K. Kanagalingam, R.E. Kouri*

Division of Toxicology and Oncology, Microbiological Associates, Bethesda, Md., USA

The human health effects of exposure to chemicals, both alone and in conjunction with other environmental and occupational agents, continue to be of major public concern. Today, some 30 government agencies and many industrial and research organizations are actively working to identify and characterize potentially hazardous substances and to understand the human health implications of exposure to these compounds.

A significant route of exposure for introducing toxic and other hazardous substances to the body is through the respiratory tract. Each day, adults respire 10,000–20,000 litres of air containing particulates such as occupational and atmospheric dusts, commercial aerosol products, smokes, radioactive particles, and other noxious agents. Respirable aerosols are deposited along the entire respiratory tract, and play a significant role in initiating or exacerbating a variety of pulmonary, cardiovascular, or infectious diseases [3].

The mechanism by which these agents exert such wide ranging toxicological and/or carcinogenic effects is not known. Given the complexity of the biologic response, an important aspect of experimental inhalation studies, especially in model animal systems, concerns limiting the route of exposure to inhalation only, thus avoiding dermal or oral exposures.

Traditional inhalation chamber exposures can result in a distribution pattern where only 10–30% of the particulates are deposited in the lungs of the exposed animals [6]. Most of the materials are deposited on the hides of the animals and as a result of preening, 60–80% of these materials are eventually found in the gastrointestinal tract [6]. A need for alternative exposure sys-

[1] Supported in part by contracts from the Council for Tobacco Research, USA, Inc.

tems has led to the development of the 'nose-only' or 'head-only' exposure systems. Under these conditions, animals are restrained in stock-like holders or body tubes and the nose of each animal protrudes into either a chamber or channel where the aerosols are delivered [11]. Using this exposure mode, the amount of total particulate matter deposited in the respiratory tract may be maximized with only a negligible deposition in the hides or the gastrointestinal tract.

With the availability of these 'nose-only' exposure systems, where the dose to the lung can be controlled and documented, a battery of both short-term and long-term tests can be utilized to evaluate or compare the biological activity of inhaled materials. Through studies of pharmacokinetics and metabolism, the deposition, distribution and excretion of the inhaled particles may be monitored, which in turn may provide information as to the targets of susceptibility, and nature of the active metabolites or detoxified products. Furthermore, following subchronic exposures, short-term endpoints of toxicity, such as morphological and biochemical changes in tissues, cellular proliferation, primary DNA damage (sister chromatid exchange, SCE) and alterations in immune and endocrine functions may be assessed.

Employing a nose-only inhalation system and using one particular environmental aerosol, cigarette smoke, results are discussed with respect to deposition and distribution of inhaled smoke particulates and some of the short-term toxicological effects induced by cigarette smoke.

## *Material and Methods*

*Animals.* Female BC3Fl/Cum mice (C57Bl × C3H/Anf Cumberland View Farms, Clinton, Tenn.) were purchased at 4–6 weeks of age. The health status of the animals and their housing conditions have previously been described [9]. Mice were vaccinated against Sendai virus and held in quarantine for a minimum of 3 weeks prior to treatment.

*Stock-Like Animal Holders.* A schematic of the animal containment system (Process and Instruments, Brooklyn, N.Y.) for 'nose-only' exposures is shown in fig. 1. Groups of 3 or 5 animals are held in stock-like holders using a combination of a neck slot and a restraining spring. The chin support insures that the nose of the animal is aligned with the cone-shaped opening on the exposure module or chamber. The nose of each animal passes through a dental rubber dam diaphragm forming a seal that prevents exposure of the body to the aerosol.

*Cylindrical Tube Animal Holders.* Animals are held in tapered cylindrical polycarbonate tubes that allow the nose of the animals to protrude into the exposure chamber. Animals are held in place by a soft sponge plug. This system has been described in detail elsewhere [7, 11]. 20 mice, 12 hamsters, or 6 or 12 rats can be simultaneously exposed using this restraint system.

*Cigarette Smoke Generation.* Fresh whole smoke from Kentucky reference 2Rl or 3Al cigarettes (University of Kentucky, Lexington, Ky.) was generated using the dynamic, large

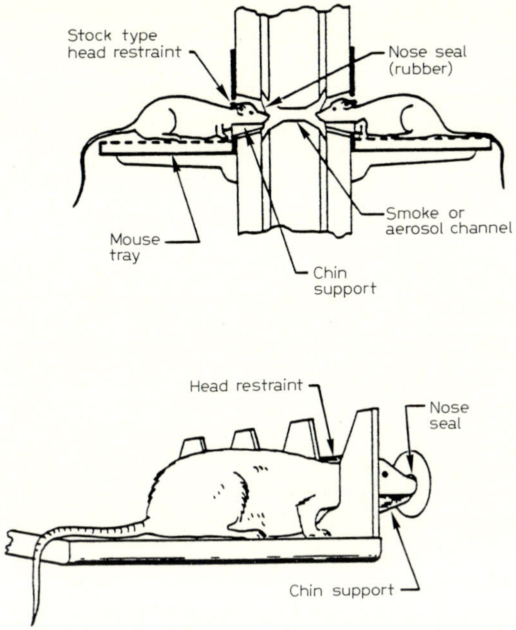


*Fig. 1.* Schematic diagram of stock-like restraints and exposure units. Cross section of the exposure unit is shown for dynamic exposures where the aerosol is delivered in a flowing stream through a smoke or aerosol channel.

capacity SEM II smoking machine [10] (Process & Instruments, Brooklyn, N.Y.), or the smaller capacity Walton Horizontal Smoking Machine (Process & Instruments). With the SEM II, the smoke aerosol is delivered in a rapidly flowing stream to animals restrained in the stock-like holders. Up to 60 mice can be exposed to aerosolized material on one exposure module and up to 8 modules have been utilized for simultaneous exposure to one aerosol. The features of the SEM II have been described in detail [10, 12, 17]. With the Walton, the smoke aerosol is generated directly into an exposure chamber. Animals restrained in either the stock-like holders or the cylindrical tubes respire directly into the chamber. The features of the Walton have also been previously described [7, 11].

*Standard Conditions for Smoke Exposure.* Smoking machines approximate human smoking by using standard smoking conditions which are the average values obtained from the habit of a large number of smokers [19]. A standardized smoking regimen has been established where the cigarette is puffed once per minute over a 2-second period, an average puff volume of 35 ml is maintained, and the cigarette is smoked to a 23-mm butt length (for non-filtered cigarettes).

*Short-Term Endpoints.* Pharmacokinetic techniques of analyzing tissues from animals exposed to tracer-labeled aerosolized material have been described in detail [10]. Cellular proliferation assays using DNA labeling index were performed according to procedures in footnotes to table IV. SCE analysis in bone marrow cells has also been described elsewhere [1, 2].

*Table I.* Distribution of $^{99m}$Tc at the conclusion of a whole-body aerosol exposure [modified from ref. 6][1]

| Tissue | nCi | % of total nCi internal | % of total nCi total animal |
|---|---|---|---|
| Lung | 622 (0.17)[2] | 23.6 | 12.8 (0.16) |
| Trachea | 12 (0.75) | 0.5 | 0.2 (1.00) |
| Head | 265 (0.21) | 10.1 | 5.5 (0.24) |
| Gastrointestinal tract | 1,381 (0.34) | 52.4 | 28.5 (0.28) |
| Carcass | 353 (0.52) | 13.4 | 7.3 (0.52) |
| Total internal | 2,633 | – | 54.4 |
| Pelt | 2,207 (0.28) | – | 45.6 (0.13) |

[1] 11 Fischer 344 rats were evaluated after 130 min exposure in a multitiered exposure chamber to a cesium chloride aerosol labeled with radioactive $^{99m}$Tc. The activity median aerodynamic diameter (AMAD) and the geometric standard deviation of the aerosol were 2.1 and 1.7 μm, respectively. The aerosol concentration was ~ 180 nCi.
[2] Coefficient of variation.

## *Results and Discussion*

Distribution pattern of $^{99m}$Tc in rodents exposed to the chemical by the chamber method is given in table I [6]. It is evident that chamber inhalation exposures to respirable particulate aerosols result in minimal dose reaching the lung (~13% of the total dose after 130 min exposure). Most of the material was deposited on the hides of the animals during the exposure and as a result of preening, it is estimated that 60–80% of these materials could eventually be found in the gastrointestinal tract [6].

The two animal restraint systems (stock-like holders and cylindrical tubes) described in Methods have been used extensively in our laboratories to expose mice 'nose-only' to a variety of aerosolized materials. For example, the deposition and distribution pattern of $^{14}$C-dotriacontane (DTC) labeled cigarette smoke particulates were compared in mice held by these two restraint systems, and simultaneously exposed to smoke generated on the Walton machine. DTC is a relatively inert particulate constituent of cigarette smoke. $^{14}$C-DTC can be added to cigarettes and recovered in the particulate phase after burning. As such, it is used to quantitate deposition and distribution of smoke particulates [for more detailed discussion see ref. 10, 11]. As shown in table II, use of tubes or neck-holders were compared in mice held

*Table II.* Percent distribution of $^{14}$C-dotriacontane (DTC) from cigarette smoke in mice exposed in whole body tubes or stock-like holders[1]

| Tissue | % of total dpm in internal tissues[2] | | % of total dpm in whole mouse[3] | |
|---|---|---|---|---|
| | tubes | holders | tubes | holders |
| Lung | 73 | 74 | 45 | 62 |
| Larynx | 2 | 4 | 1 | 3 |
| Head | 7 | 6 | 4 | 5 |
| Stomach | 13 | 11 | 8 | 9 |
| Other internal[4] | 6 | 5 | 4 | 4 |
| Total internal | – | – | 62 | 83 |
| Hides[5] | – | – | 38 | 17 |

[1] Smoke exposures were carried out on the Walton Horizontal Smoking Machine under standard exposure conditions, using 10% $^{14}$C-DTC-2A1 cigarette smoke for 30 s of smoke alternating with 30 s of air, each minute for 10 consecutive minutes. Mice were sacrificed immediately after exposure and selected tissues analyzed for radioactivity. Data are the means from 12 mice. Mice restrained in both holders and tubes were exposed to the same smoke samples: the tubes were on one side of the chamber and the neck-holders were on the other side.
[2] The total amount of radioactivity in all internal tissues was determined, excluding hides, and the percent given.
[3] The total body burden was determined from the amount found in the internal tissues and the hides.
[4] Consisted of a composite sample of all other internal tissue and the carcass.
[5] Consisted of the skin and fur as a composite sample.

by these two restraint systems, and simultaneously exposed to smoke generated on the Walton machine. The use of either holder for restraint substantially increased the dose to the lung and decreased the dose to the pelt when compared to the chamber-exposure method (table I). Using the cylindrical tube restraint, the dose of material to the lung was increased ~3-fold over the whole body exposure (table I), while using the neck-holder restraints, the dose was increased ~5-fold. In both cases, the dose to the gastrointestinal tract was reduced ~3-fold. If an assumption is made that during 24 h postexposure, approximately 80% of the material on the pelt is ingested through preening, then the whole body exposure would result in 70% of the material in the gastrointestinal tract, compared to 44% for the tube-holders and only 24% for the neck-holders. It should be noted that the absolute dose to the lung is the same for the cylindrical tubes and the neck holders – in either system 73–74% of the total internal dose is found in the lung. In contrast, only 24% of the total internal dose may reach the lung when certain whole-body chamber exposures are performed (table I).

A battery of short-term end points were then applied to assess the biological effects of materials which are introduced into the body via this inhalation system. Three assays will be discussed: a pharmacokinetics assay, a cellular proliferation assay and a cytogenetics assay. These assays were performed on mice exposed 'nose-only' to whole cigarette smoke generated on either the SEM II or Walton machine. The deposition and distribution of the smoke particulates generated using either of these smoking machines has been shown to be equivalent for the same aerosol concentration and exposure times [11].

A pharmacokinetics assay was carried out to determine the fate of catechol inhaled in cigarette smoke. Catechol is a major component of cigarette smoke (up to 0.25 mg/cigarette) [4] and has recently been shown to have cocarcinogenic properties on mouse skin [8]. $^{3}H$-labeled catechol was synthesized [13] and injected into cigarettes, from which it could be recovered in the particulate phase after burning. Mice were exposed for 10 min to $^{3}H$-catechol-labeled cigarette smoke generated on the Walton under standard smoke exposures conditions, sacrificed and tissues analyzed for radiolabeled material. Immediately after the 10-min exposure ('zero time'), over 55% of the radioactivity was found in the blood, ~32% in the internal tissues, and less than 12% in the respiratory tract [13]. In addition, the total amount of material recovered from the tissues of the exposed animals decreased rapidly within 2 h. Analysis of urine and feces collected for 2 h after exposure showed that 94% of the radioactivity was in the urine (table III). Less than 0.5% of the total dose was found in the lung at this time. These data are consistent with the scheme that catechol is initially deposited in the lungs of smoke-exposed mice and is rapidly redistributed to the blood and tissues (especially liver). This process has a half-life of only 10–15 min from the first inhalation of $^{3}H$-catechol-labeled cigarette smoke [13].

The pharmacokinetic study demonstrated that exposure to cigarette smoke could result in certain smoke components exerting effects directly in the lung (e.g., DTC) and other chemicals exerting effects systemically (e.g., catechol). Two assays were selected which showed promise in investigating the potential direct pulmonary effects and systemic activity of inhaled cigarette smoke. DNA labeling index (a measure of cellular proliferation) in the lung, was used as an indicator of localized smoke-induced effects and SCE studies in bone marrow were employed as an indicator of systemic

*Table III.* Distribution of $^{3}H$-labelled radioactivity (dpm) in tissues and excreta of BC3Fl/Cum mice 120 min after exposure to 10% (v/v) 2R1 smoke labeled with $^{3}H$-catechol[1] [from ref. 10]

| Tissue or site | dpm/organ/mouse | % of total dpm |
|---|---|---|
| Lung | 37.5 (0.45)[2] | 0.5 |
| Turbinate | 12 (2.00) | 0.2 |
| Total respiratory tract[3] | 50 | 0.7 |
| Blood | 133 (0.47) | 2.0 |
| Liver | 33 (0.36) | 0.5 |
| Kidney | 17 (0.04) | 0.2 |
| Stomach | 8 (1.16) | 0.1 |
| Remaining tissues | 12.5 | 0.2 |
| Total internal tissues | 254 | 3.6 |
| Urine[4] | 6,375 | 91.2 |
| Feces[4] | 108 | 1.5 |
| Total body | 6,990 | |

[1] 12 BC3Fl/Cum female mice were exposed to 10% (v/v) $^{3}H$-catechol-2R1 cigarette smoke generated on Walton under standard exposure conditions for 300 s total smoke exposure time. After exposure, mice were immediately placed in a metabolic cage to collect urine and feces. After 2 h, mice were sacrificed by $CO_2$ asphyxiation.

[2] Coefficient of variation.

[3] Represents the sum of the dpm in lung, turbinate, larynx and trachea.

[4] Urine and feces were pooled samples from 12 mice and data presented on an individual mouse basis.

effects. Data are presented in table IV for animals restrained in the neck-holders and exposed daily by 'nose-only' inhalation to smoke generated on the SEM II from 2Rl or 3Al cigarettes and compared with sham-exposed or untreated animals.

When animals were pulsed for a 1-hour period with $^{3}H$-thymidine immediately after smoke exposure and the pulmonary tissue analyzed autoradiographically, DNA synthesis was found to have increased 3- to 6-fold compared to controls after 4 and 12 weeks of daily smoke exposure (table IVa). Labeling was confined to the cells lining the alveoli; none was found in the cells of the bronchial epithelium. Analysis of the bone marrow of mice also revealed that after 4 or 12 weeks exposure to smoke from either cigarette type, an ~2-fold increase in the number of SCEs/cell was observed compared to the controls (table IVb). Analysis of the grain counts from the autoradiograms suggests that some sort of 'DNA repair' synthesis may be

*Table IV.* Lung DNA labeling index (LI) and SCE in bone marrow of mice exposed to whole cigarette smoke[1]

| Exposure period[2] | Treatment group | | | |
|---|---|---|---|---|
| | 3A1 | 2R1 | sham[3] | shelf |
| *a. Percent labelled cells* | | | | |
| Lung DNA LI[4] | | | | |
| 3– 4 weeks | 1.15[5] | 1.16[5] | 0.23 | 0.16 |
| 12–13 weeks | 0.88[5] | 0.62[5] | 0.23 | 0.16 |
| *b. Average number SCE/metaphase* | | | | |
| SCE in bone marrow[6] | | | | |
| 3– 4 weeks | 8.84[5] | 7.57[5] | 3.79 | 3.92 |
| 12–13 weeks | 8.38[5] | 8.42[5] | 3.96 | 4.47 |

[1] Exposure consisted of 10% (v/v), ~ 9 μg TPM/ml) 2R1 or 3A1 cigarette smoke under conditions of 15 s of smoke, followed by 45 s of air each minute for 126 consecutive minutes.
[2] Exposures were performed daily 5 days/week.
[3] Sham-exposure controls consisted of age-matched mice, subjected to the same manipulations as the smoke-exposed mice, including sham exposure on a smoking machine without cigarettes.
[4] For each exposure period $^3$H-thymidine was administered intraperitoneally immediately following final smoke exposure, animals sacrificed 1 h later, lungs fixed, sectioned and processed for autoradiography. The number of cells labelled among a total of 4,000–6,000 cells counted was determined and is presented as the percent of total cells labeled. Data are from 4 mice per exposure group (3A1, 2R1, and sham), and 2 shelf controls.
[5] Significantly different from sham control, t test, $p \leq 0.05$.
[6] At least 50 s replication-cycle metaphase cells were analyzed according to *Benedict* et al. [2]. Data are presented as the mean of 2–3 animals/group.

occurring [14, 15]. Thus, the results from two assays suggest cigarette smoke contains chemicals which may cause DNA damage, not only in the pulmonary cells at the site of the original deposition, but also in cells at sites remote from the lung (e.g., bone marrow).

When the deposition pattern and pharmacokinetics of a specific chemical are understood, then a variety of specific end points may be monitored. For example, 'nose-only' exposure to whole cigarette smoke, in addition to the aforementioned effects, also results in changes in the following endpoints: (a) induction of pulmonary microsomal monooxygenase activity [16], (b) induction of pulmonary ornithine decarboxylase activity [5], and (c) inhibition of pulmonary DNA repair capacity [18]. Thus, a well-defined exposure system where greater than 90% of the aerosol is deposited in the respiratory tract of the exposed animals can now be used concomitantly with a variety of standard short-term (usually less than 14 days) end points. This approach will provide a more effective and efficient method for evaluating aerosols of potential biological activity.

## *References*


1 Benedict, W.F.; Banerjee, A.; Kanagalingam, K.K.; Dansie, D.R.; Kouri, R.E.; Henry, C.J.: Increased sister chromatid exchange in bone marrow of mice exposed to whole cigarette smoke. Carcinogenesis (in press, 1982).

2 Benedict, W.F.; Banerjee, A.; Venkatesan, N.: Cyclophosphamide induced oncogenic transformation, chromosomal breakage and sister chromatid exchange following microsomal activation. Cancer Res. *38:* 2922–2924 (1978).

3 Brain, J.D.; Valberg, P.A.: Deposition of aerosol in the respiratory tract; in Murray, Lung disease state of the art, pp. 225–274 (American Lung Association, New York 1978/1979).

4 Brunnemann, K.D.; Lee, H.C.; Hoffmann, D.: Chemical studies on tobacco smoke. XLVII. On the quantitative analysis of catechols and their reduction. Analyt. Lett. *939:* 945 (1976).

5 Dinowitz, M.; Nims, R.; Bhooshan, B.; Kouri, R.E.; Henry, C.J.: Induction of ornithine decarboxylase (ODC) by 12-0-tetradeconoylphorbol-13-acetate (TPA) in pulmonary tissue: a model system for tumor promotion in mouse lungs. Proc. Am. Ass. Cancer Res. *21:* 31 (1980).

6 Griffis, L.C.; Wolff, R.K.; Beethe, R.L.; Hobbs, C.H.; McClellan, R.O.: Pulmonary deposition of a $^{99m}$Tc labeled aerosol in a whole-body exposure; in Inhalation toxicology research institute annual report, pp. 259–265 (Lovelace Biomedical Environmental Res. Inst., Albuquerque 1979).

7 Guérin, M.R.; Stokely, J.R.; Higgins, C.E.; Moneyhun, J.H.; Holmberg, R.W.: Inhalation bioassay chemistry – Walton horizontal smoking machine for inhalation exposure of rodents to cigarette smoke. J. natn. Cancer Inst. *63:* 441–448 (1979).

8 Hecht, S.S.; Carmella, S.; Mori, H.; Hoffmann, D.: A study of tobacco carcinogen. XX. Role of catechol as a major cocarcinogen in the weakly acidic fraction of smoke condensate. J. natn. Cancer Inst. *66:* 163–169 (1981).

9 Henry, C.J.; Billups, L.H.; Avery, MD.; Rude, T.H.; Dansie, D.R.; Lopez, A.; Sass, B.: Whitmire, C.E.; Kouri, R.E.: Lung cancer model system using 3-methylchol-anthrene in inbred strains of mice. Cancer Res. *41:* 5027–5032 (1981).

10 Henry, C.J.; Caton, J.E.; Stokely, J.R.; Guérin, M.R.; Lopez, A.; Avery, M.D.; Dansie, D.R.; Henderson, G.M.; Gayle, T.; Whitmire, C.E.; Kouri, R.E.: Deposition and distribution of the total particulate matter of cigarette smoke in mice using a large capacity smoke exposure system. Toxicol. appl. Pharmacol. *58:* 399–409 (1981).

11 Henry, C.J.; Whitmire, C.E.; Lopez, A.; Dansie, D.R.; Avery, M.D.; Caton, J.E.; Stokely, J.R.; Holmberg, R.W.; Guérin, M.R.; Kouri, R.E.: The dosimetry and distribution of whole cigarette smoke particulates in inbred strains of mice: comparison of a large smoke-exposure machine (SEM) with a small-capacity smoke-exposure machine (Walton); in Sanders, Cross, Dagle, Mahaffey, Pulmonary toxicology of pulmonary toxicology

of respirable particles, pp. 177–192 (Technical Information Center, US Department of Energy, NTIS, 1980).
12 Holmberg, R.W.: Determination of particle size in tobacco smoke in inhalation exposure devices using methylcyano-acrylate fixation and electron microscopy; in Guerin, Stokely, Higgins, Tobacco smoke inhalation bioassay chemistry, pp. 103–118 (DOE Report ORNL-5424, Oak Ridge National Laboratory, NTIS, Oak Ridge 1979).
13 Hwang, K.K.; Sonko, C.; Dansie, D.R.; Kouri, R.E.; Henry, C.J.: Studies on the deposition and distribution of catechol from whole cigarette smoke in BC3Fl/Cum mice. Toxicol. appl. Pharmacol. (in press, 1982).
14 Benedict, W.F.; Kouri, R.E.; Henry, C.J.: Cigarette smoke increases DNA replication in lung and sister chromatid exchange (SCE) in bone marrow of mice. Am. Ass. Cancer Res. *23:* 56 (1982).
15 Kanagalingam, K.K.; Reed, S.M.; Dansie, D.R.; Hall, W.C.; Kouri, R.E.; Henry, C.J.: Autoradiographic analysis of DNA synthesis in pulmonary tissues of mice exposed to whole cigarette smoke. Toxicologist *2:* 46 (1982).
16 Kouri, R.E.; Rude, T.; Curren, R.D.; Brandt, K.R.; Sosnowski, R.G.; Schechtman, L.M.; Benedict, W.F.; Henry, C.J.: Biological activity of tobacco smoke and tobacco smoke related chemicals. Environ. Health Perspect. *29:* 63–69 (1979).
17 Moneyhun, J.H.; Stokely, J.R.; Florant, L.: Process and Instruments Corporation automatic smoke exposure machine SEM II; in Guérin, Stokely, Higgins, Tobacco smoke inhalation bioassay chemistry, pp. 19–33 (DOE Report ORNL-5424, Oak Ridge National Laboratory, NTIS, Oak Ridge 1979).
18 Rasmussen, R.E.; Boyd, C.H.; Dansie, D.R.; Kouri, R.E.; Henry, C.J.: DNA replication and unscheduled DNA synthesis in lungs of mice exposed to cigarette smoke. Cancer Res. *41:* 2583–2588 (1981).
19 Wynder, E.; Hoffman, D.: Smoking machines and smoking techniques; in Tobacco and tobacco smoke, pp. 94–120 (Academic Press, New York 1967).

Dr. C.J. Henry, Division of Toxicology and Oncology, Microbiological Associates, 5221 River Road, Bethesda, MD 20816 (USA)

Safety Evaluation and Regulation of Chemicals. 1st Int. Conf., Boston, Mass., 1982, pp. 243–250 (Karger, Basel 1983)

# Detection of Rare Adverse Reactions Induced by Chemicals

*Tibor Balazs*

Division of Drug Biology, Food and Drug Administration, Washington, D.C., USA

Unexpected toxic effects may occur in humans after exposure to chemicals that appeared to be innocuous from toxicity studies in experimental animals. Similarly, adverse effects of drugs may develop in humans given therapeutic doses regardless of whether the effect occurred even at the highest doses in preclinical studies or initial clinical trials.

The problem of the lack of detection of these effects in preclinical studies is inherent in the species used and the scope of examinations. The unpredictability of these effects from the initial clinical trials is attributable to their low incidence in humans. Most of these effects have been termed idiosyncratic. Such a characterization will become obsolete through understanding of the mechanisms of action of the drug, which will provide the tool for detection of the effects in preclinical or early clinical studies.

The effects that are not detected in premarketing or preclinical safety evaluation studies can be divided into two categories from the vantage point of the experimental toxicologist. In the first category, the biochemical mechanisms underlying the toxicological effects occurring only in humans are present or can be induced in experimental animals. Their detection or their induction, however, requires unconventional approaches. The second category includes those effects that require predisposing conditions which may or may not be obtainable in experimental animals; the best-recognized conditions are the genetic makeup and physiological state of the individual. Examples of rare adverse effects and approaches for their detection are presented.

*Genetic Influence on the Occurrence of Rare Reactions*

The mechanisms of toxic injuries often involve the formation of reactive metabolites from the parent compound; these include such metabolites as free radicals, epoxides, and carbonium ions, which interact covalently in situ with structural lipids, proteins, or nucleic acids and initiate cellular injury. Various endogenous substances, e.g., free radical scavengers, can prevent this interaction because of their high chemical affinities for the reactive metabolites. If the free radical scavenger is not present in a quantity sufficient to compete successfully with the neighboring structural molecules, the toxic reaction occurs. Thus, the development of injury depends upon the amounts of toxic metabolites formed and the availability of protective substances. Both of these are subject to species and individual variations, which can handicap the prediction of potential toxicity from one species to another. This phenomenon also provides an explanation for the unexpected sensitivity of some individuals within a species; hence genetic makeup can be responsible for rare toxic reactions.

An example is the action of acetaminophen, which is a weak hepatotoxin in most species. In experimental animals, its hepatotoxicity can be increased by pretreatment with an inducer of hepatic drug-metabolizing enzyme systems, for example, with phenobarbital or a depletor of glutathione such as diethylmaleate. This procedure has been used to detect potential hepatotoxicity of a number of chemicals that are converted to reactive metabolites in the liver [9].

In some instances, additional predisposing factors which are encountered in the use of the chemical are required for the induction of hepatotoxicity. For example, severe liver necrosis occurred in about 1 of 7,000 patients given halothane. However, when this anesthetic was tested in rats pretreated with phenobarbital the effect could be detected in only one strain, the Fischer rat. In another strain, the inhaled oxygen concentration had to be reduced in addition to phenobarbital pretreatment to elicit the lesion. Pretreatment with Aroclor 1254 (a polychlorinated biphenyl) instead of phenobarbital also elicited the lesion in rats of the latter strain. These data are consistent with the concept that the scope of enzyme induction varies with the inducer, strain of animal, and experimental conditions, since stimulation of a reductive pathway is necessary for formation of the hepatotoxic metabolite of halothane [7].

A number of reactive metabolites may be generated from a single compound, resulting in a variety of toxic effects. For example, the incidence of isoniazid-induced liver injury is greater in fast acetylators, but the incidence of isoniazid-induced neuropathy or autoimmune disease is greater in slow acetylators, since a nonacetylated metabolite is responsible for the latter reactions. In accord with this concept, pretreatment with phenobarbital enhanced the hepatotoxic effect of isoniazid in rats, and drug-induced autoimmune reactions have occurred in strains of mice that are genetically slow acetylators [8, 12].

Although induction of drug-metabolizing enzymes can also be elicited in organ systems other than the liver, it has not been extensively used for the detection of potential extrahepatic toxic effects. However, enzyme induction has been successfully used for enhancement of the renal and pulmonary toxicity of xenobiotics [8].

Genetic difference, which is the basis of a number of rare toxic reactions, has also been recognized as a factor in the site of action of chemicals. The lack of identical conditions in experimental animals has been a problem in the detection of such reactions. For example, a deficiency of erythrocyte glucose 6-phosphate dehydrogenase activity results in the deficiency of reduced glutathione. This condition renders the cell vulnerable to hemolytic effects of primaquine-type antimalarial drugs and several sulfonamides, as well as other compounds. Although enzymes of low activity may be found in some species of animals, there is no appropriate animal model. Thus the screening of volunteers in cautious clinical trials for enzyme deficiencies and their consequences appears to be in order. However, there are some inborn metabolic abnormalities in animals that resemble those occurring in humans. Malignant hyperthermia with rigidity, which is caused by a few general anesthetics, is brought about in humans by an excessive contraction of skeletal muscle due to abnormal calcium fluxes. Pigs of a specific strain have been found to be susceptible to this effect.

Some chemicals can also produce degenerative and inflammatory effects via the immune system. These chemicals combine with proteins or nucleic acid and act as antigens to initiate a variety of immune reactions. The mechanisms of several hepatic reactions caused by drugs (iproniazid, aminosalicylate, sulfonamide, and nitrofurantoin) are in this category, since they are often associated with skin rashes, blood dyscrasias, and eosinophilia, and a later challenge is often followed by a recurrence of symptoms. Drug-induced blood dyscrasias such as methyldopa-induced hemolytic anemia and thrombocytopenia have been shown to be mediated by the immune system (Coombs-positive reactions); similarly, gold salts, penicillamine, and sulfonamide induced glomerulonephritis, as shown by the presence of their immune complexes. These reactions hitherto have not been detected in experimental animals. The incidence of these reactions is very low, which is attributable to the great variability in genetic determinants of the immune response, i.e., the major histocompatibility or immunogenetic complex. Although only a few of the above-mentioned reactions (nitrofurantoin hepatitis, gold salt nephritis, and hydralazine-induced autoimmune disease) have been associated with a specific haplotype of the human major histocompatibility complex, it is conceivable that susceptibility to other drug-induced immune reactions is also linked to a specific haplotype. In a few instances such a relationship has been detected in experimental animals. Genetically determined reactivity to mercuric chloride-induced immune complex glomerulonephritis has been demonstrated in rats. Brown Norway rats bearing a specific haplotype ($RTl^n$) are sensitive; other strains (bearing the $RTl^1$ haplotype) are resistant, but high doses of mercuric chloride overcome the resistance in one of these strains [4].

Utilization of rats with various RTl or mice with various $H_2$ histocompatibility complexes may provide a tool for selecting the correct haplotype for prediction of adverse drug effects that are conditional upon the presence of specific immune response genes.

### *The Effect of Physiological, Pharmacological, and Other Factors on Development of Rare Reactions*

Newborns are at great risk from xenobiotics because of their low rate of detoxifying and excretory functions. Chloramphenicol-induced cardiovascular collapse and aminoglycoside antibiotic-induced deafness in infants and neonatal animals are explained on this basis. The low level of enzyme activity at the target site is responsible for some of the unique sensitivity, e.g., sensitivity of the neonatal erythrocyte to the hemolytic effect of oxidizing agents. The lack of a blood-brain barrier is responsible for the encephalopathy caused by cerebral penetration of bilirubin that is displaced from serum albumin by certain chemicals. A few chemicals, when given at a critical period, interfere with the postnatal development of a specific system, resulting in functional defects in the central and autonomic nervous system or in the differentiation of the neuroendocrine and immune systems in experimental animals [1]. Some of these alterations are detected later in life when the function of the affected system is operative, e.g., the reproductive system. In people, the appearance of an abnormality that is detected at times remote from the time of exposure is difficult to relate to the delayed effect of xenobiotics. For example, gout in the aging has been said to be a consequence of excessive lead exposure in childhood [6]. The experience with diethylstilbestrol as a transplacental carcinogen in experimental animals and in humans was of heuristic significance for demonstrating a delayed rare neoplastic reaction. These examples suggest that chemicals destined for use during pregnancy or in children should be tested for neonatal toxicity and for delayed effects.

The incidence of drug-induced toxic reactions increases greatly during senescence. This is attributed to the decreased rate of drug metabolism and excretion as well as to changes in homeostatic capability. Paradoxical effects of drugs that act on several neuroreceptors may occur because of losses in some of the receptors with age [2]. A decline in the function of the immune system is probably responsible for the higher incidence of drug-induced autoimmune reactions in elderly patients. In old age, defects in the functions of T lymphocytes are detectable. Humoral factors, e.g., 'facteur thymique sérique' regulating immune response, decline with advancing age in animals and humans. It would be advantageous to use aged animals for preclinical toxicology of geriatric products; however, the selected animals should be free from pathological changes caused by intercurrent diseases.

Rare toxic effects may not be attributable to a single agent, as shown by numerous examples of drug interaction. The chance of a severe reaction is particularly great with the concurrent use of two drugs that have similar pharmacological effects occurring by different mechanisms. Physicians may be confronted with situations, e.g., lack of sufficient therapeutic response of a drug, in which they must use another drug concurrently, even though the safety of the combination has not been established. A recent example illustrates this problem. Severe cardiodepressant effects occurred with the combined use of beta-adrenergic blockers and calcium channel blockers in patients with beta blocker-resistant angina. Subsequent experiments demonstrated the marked potentiation of the median lethal dose of propranolol in mice pretreated with verapamil. Atrioventricular block and profound hypotension preceding death occurred in rabbits treated with low multiples of pharmacological doses of the combination [*Balazs, Kandil, and Vick,* unpublished data].

Interactions of drugs with food ingredients can also lead to adverse reactions. One of the best-known examples is the hypertensive crisis brought about in humans by the interactions of monoamine oxidase inhibitor antidepressant drugs and tyramine-containing food. This effect has also been shown in animal experiments.

Adverse interactions of environmental and industrial chemicals have also been recognized; examples are the synergistic pulmonary effects of certain particulates with sulfur dioxide and the increased incidence of cancer in asbestos workers who are cigarette smokers. The development of cancer best demonstrates the role of chemical interaction inasmuch as two chemicals, an initiator and a promoter, are often required in chemical carcinogenesis.

It is difficult to foresee all the conditions, including the chemical environment, that play a role by a specific interaction in eliciting rare toxic effects. The commercial use of a product may aid in the selection of chemicals or animal models for studies of this nature. A good example is the cardiomyopathy that occurred in heavy beer drinkers; it was induced by a cobalt salt used in beer as a foam stabilizer. The cardiomyopathy could be best reproduced in animals maintained solely on beer and protein- or thiamine-deficient diets [11].

During long-term administration of drugs that act on the central or autonomic nervous system, adaptation involving changes in the receptors can develop. This phenomenon can result in adverse reactions that may develop during or after treatment. The biochemical changes responsible for the clinical effects are also present in experimental animals. Their detection, however, in the absence of clinical signs, requires special pharmacological testing.

An example is tardive dyskinesia, a chronic movement disorder which develops in about 10% of patients on long-term therapy with neuroleptic drugs. Supersensitivity of dopamine receptors plays a part in the pathomechanism of the disorder. Although the abnormal movements are not observed in experimental animals, the presence of receptor supersensitivity can be demonstrated after withdrawal of the agents that produce the dyskinesia in humans. This supersensitivity is detected by an exaggerated response to a dopaminergic agonist [3].

A homologous phenomenon occurs after the withdrawal of beta-adrenergic receptor-blocking drugs. Rebound effects have been observed in patients with angina or hypertension and can result in myocardial infarction. The number of beta receptor binding sites in experimental animals increases during treatment. This increase most likely also occurs in humans and could be responsible in part for the withdrawal syndrome, i.e., the exaggerated effects of endogenous catecholamines. An increased pharmacological reaction to adrenergic drugs has been measured in normal volunteers as well as in experimental animals after withdrawal of a beta blocker [5].

The role of preexisting disease in addition to the developing receptor supersensitivity should be considered in the etiology of both of the syndromes described above. Nevertheless, the drugs are responsible for precipitation of the syndromes. These examples also illustrate the complex, often multifactorial, etiology of rare toxic reactions. It has been shown that certain toxic effects of a specific drug might become manifest during one disease but not during another, although the drug is curative for both [10].

With the wide use of chemicals and with more variables in their usage, the chance increases that some of the rare reactions may become widespread. One of the best examples is a neuropathy attributed to the use of a gastrointestinal antiseptic, a chlorohydroxyquinoline, in Japan. The drug has been used for decades throughout the world and particularly in Europe, with very few related reactions. The extensive use of the drug in Japan in aged, chronically ill patients or after major surgery, with the belief of absolute safety, caused an iatrogenic disease of epidemic proportion. The toxic effects were later reproduced in dogs with doses many times higher than those given to humans. This example emphasizes the need for continuous feedback from clinical or epidemiological data to the experimental toxicologist. The toxicologist, having evaluated these reports, studies in experimental animals the cause-effect relationships as well the significance of predisposing factors. The recognition of a specific rare toxic effect induced by the chemical provides essential information on how it can be used safely.

## *References*


1 Balazs, T.: The delayed long-term effects of chemicals following neonatal exposure in laboratory animals. Fd. Cosmet. Toxicol. *19:* 533–537 (1981).

2 Bender, A.D.: Pharmacodynamic principles of drug therapy in the aged. J. Am. Geriat. Soc. *22:* 296–303 (1974).

3 Burki, H.R.: Extrapyramidal side-effects; in Zbinden, Gross, Pharmacological methods in toxicology, pp. 525–534 (Pergamon Press, Oxford 1979).

4 Druet, P.; Sapin, C.; Druet, E.; Hirsch, F.; Bellon, B.: Nephrotoxin induced changes in kidney immunobiology with special reference to mercury-induced glomerulonephritis. Proc. Int. Symp. Nephrotoxicity, University of Surrey, Guildford (in press, 1982).

5 Ehrreich, S.J.; Kandil, A.: Cardiotoxicity of beta adrenergic blocking agents; in Balazs, Cardiac toxicology, vol. 2, pp. 113–125 (CRC Press, Boca Raton 1981).

6 Foulkes, E.C.; Hammond, P.B.: Toxicology of the kidney; in Casarette, Doull, Toxicology, p. 196 (Macmillan, New York 1975).

7 McLain, G.E.; Sipes, T.G.; Brown, B.R., Jr.: An animal model of halothane hepatotoxicity: role of enzyme induction and hypoxia. Anesthesiology *51:* 321–326 (1969).

8 Mitchell, J.R.; Corcoran, G.B.; Hughes, H.; Lauterburg, B.H.; Smith, C.V.: Acute lethal injury caused by chemically reactive metabolites; in Brown, Davies, Organ directed toxicity, pp. 117–129 (Pergamon Press, Oxford 1981).

9 Mitchell, J.R.; Nelson, S.D.; Thorgeirsson, S.S.; McMurtry, R.J.; Dybing, E.: Metabolic activation: biochemical basis for many drug-induced liver injuries; in Popper, Schaffner, Progress in liver diseases, vol. 5, pp. 259–279 (Grune & Stratton, New York 1976).

10 Raskova, H.; Eli, J.; Perlik, F.; Polanski, T.; Slovick, H.: Unexpected side effects of 6-azo uridine in rheumatoid arthritis and feedback in animal experiments; in Proc. Eur. Study Drug Toxic. XII, Int. Congr. Ser., vol. 220, pp. 191–198 (Excerpta Medica, Amsterdam 1971).

11 Rona, G.; Chappel, C.I.: Pathogenesis and pathology of cobalt cardiomyopathy; in Bajusz, Rona, Recent advances in cardiac structure and metabolism, pp. 407–422 (University Park Press, Baltimore 1973).

12 Ten Veen, J.H.; Feltkamp, T.E.W.: Studies on drug-induced lupus erythematosus in mice. Clin. exp. Immunol. *11:* 265–276 (1972).

Dr. Tibor Balazs, Bureau of Drugs, Food and Drug Administration, 200 C Street, SW, Washington, DC 20204 (USA)

Safety Evaluation and Regulation of Chemicals. 1st Int. Conf., Boston, Mass., 1982, pp. 251–258 (Karger, Basel 1983)

# Tumor Markers in Experimental Animals

*G. Mazué, D. Gouy, B. Remandet*

Centre de Recherches, Clin Midy Research Center, Montpellier, France

Classic carcinogenicity studies are both time- and money-consuming; hence the interest of having tests for early detection or, at least, for orientation at our disposal. Reliable tests based on mutagenesis now exist for the detection of genotoxic carcinogenic agents. There remain the epigenetic carcinogenic agents, such as promoters.

Amongst these tests, aimed at revealing potential promoting activity, the search for altered foci and early neoplastic liver lesions is widely used and classic histochemical markers are needed, those which are resistant to iron loading after induced siderosis.

The aim of this paper is to define what a tumor marker is and then to try to prove the usefulness of gamma-glutamyl-transpeptidase (GGT) as a reliable marker to show the promoting action of some compounds on the liver.

## *Generalities on Carcinogen Markers*

### *Definition*

A tumor marker should be a parameter easy to explore, reliable and representative of a cellular change consistent with the development of a carcinogenic phenomenon. Furthermore, there should be a good correlation between the parameter selected and the cellular changes responsible for the tumoral process.

To facilitate exploration, the parameters must be measured on samples easy to take from experimental animals, and the dosing technique must be easy, quick and reproducible in order to obtain results in the early stage of the study. The quality of the marker will be judged by its perfect correlation with the histologic lesions found – known criteria enabling the definition of carcinogenesis – as well as by a good correlation in the variations of the marker with regard to cells, tissues and biological liquids.

*Specificity of the Marker*

A carcinogen marker may be specific of either cellular dedifferentiation or dedifferentiation of a particular organ and, when possible, it should even have both specificities. Indeed, the marker used must reflect the carcinogenic process. Therefore, there should not be any major interference with other processes such as toxic cellular damage, enzymatic induction or congenital enzymatic deficit.

In order to reflect carcinogenesis well, the marker must be a consequence of the cellular dedifferentiation, or at least it should be related to it. In a few cases, production of a marker inside the cell may decrease owing to metabolic changes. Then it is a marker by default, quite frequent during dedifferentiation.

*Markers of the Cancerous Cell*

The cancerous phenomenon is associated with changes in metabolism and membranes damaging the cancerous cell. The metabolic alterations observed are the following:

– Increased synthesis of nucleic acids DNA and RNA accounting for increased basophilia of the cells affected, made evident by increased incorporation of 3H-thymidine.

– Reapparition of embryonal biochemical signs such as alpha-fetoprotein present in the vitellus and the fetal liver, met in the case of malignant dysembryomas and hepatic tumors. These phenomena show the dedifferentiation character of cancerous cells. The same may be said of the apparition of the carcino-embryonal antigen in the case of digestive and mammary carcinomas and other carcinomas of this type.

– Repression or hyperproduction of secretions or excretions. For instance, gamma-globulinemic hyperproductions in the case of plasmocytic leukemia, hypersecretion of acid phosphatases in the case of cancer of the prostate or mammary glands, or alkaline phosphatase in the case of bone or liver tumors, amylase in the case of pancreas cancer, etc.

– Apparition of abnormal proteic secretions often having hormonal activity, for instance, insulin and gastrin in pancreatic tumors.

– Membranous alterations observed in cancerous cells, which are most probably related to their loss of contact inhibition, are mainly the following:

(1) Changes in surface glycoproteins (with increased sialic acid concentration).

(2) Apparition of surface neoantigens not seen in the non-neoplastic cell, these antigens being often, but not always, specific of the tumor.

(3) It should be noted too that cultured tumor cells are agglutinated by lectins (concanavalin A), in contrast to non-tumor cells.

### *Main Carcinogen Markers in the Rat Liver*

#### *Histologic Criteria*

With common histological staining procedures, liver lesions were noticed according to *Stewart* et al. [7], and can be summed up as follows: foci and areas of cellular alteration (with clear, basophilic, acidophilic or mixed cells); neoplastic nodules (same cellular types and with compression of the surrounding parenchyma), and hepatocellular carcinomas (trabecular carcinoma or adenocarcinoma).

These lesions have to be interpreted carefully in view of the histochemical or histoenzymatic criteria. Different methods were used for this purpose. All the above-mentioned morphologic alterations are characterized as follows: increased glycogen storage; decreased activity of glucose-6-phosphatase (G6Pase) and adenosine triphosphatase (ATPase); apparition of GGT activity, absence of iron storage after induced hepatic siderosis; hypercolorability after staining with toluidine blue (nucleic acids). Of all these methods, the one based on the inability to fix the iron was found to be the most reliable, the others giving irregular results [2, 3].

Here appears the major disadvantage met in this type of study: the data are available only after sacrifice of the animals, followed by a protracted period for processing. This is why data should be available at an earlier stage in the study in order to appreciate the carcinogenesis process. These data are provided by biochemistry, both tissular (after grinding and homogenization) and plasmatic.

#### *Biochemical Markers [4]*

Biochemical markers may be studied in organs in the form of homogenates, as well as in plasma and urine. They are especially proteinaceous, but they may also be hormones in the case of secreting glandular tumors. Among proteins, two groups may be distinguished: those having an enzymatic activity, they are the most numerous, and those which have no enzymatic activity.

The proteins having no enzymatic activity usually correspond to the dedifferentiation phenomenon: (alpha-fetoprotein, carcino-embryonal antigen, diverse glycoproteins). In contrast, changes in enzymatic activity correspond to metabolic alterations or to cellular destruction.

The main enzymes used as markers are related to the liver and are studied in plasma: alkaline phosphatases, lactate dehydrogenase (LDH), transaminases, aldolase, phosphohexose isomerase, dehydrogenase malate, dehydrogenase isocitrate, leucine aminopeptidase (LAP), 5′-nucleotidase, glutathione reductase, GGT. Other organs can also be explored, such as the pancreas with plasmatic LAP, as well as plasmatic or urinary amylase.

A greater specificity may be obtained by using isoenzymes, i.e. proteins having the same enzymatic activity but with different unicellular structures more specific of an organ and identifiable by means of electrophoresis. The proportion of different enzymes may be changed during carcinogenesis, especially with LDH and aldolase and, possibly, also with GGT.

Among these markers, GGT presents particular interest for the study of hepatic carcinogenesis in the rat. In fact, this enzyme is virtually undetectable in the liver of the adult rat. It is localized in the cells which border the bile ducts and the plasmatic rate is almost nonexistent. But it appears in liver homogenates, as in plasma in the case of hepatic carcinogenesis. The plasmatic base rate being almost nonexistent, its increase is easily revealed. Therefore, this enzyme has an organ specificity as well as a dedifferentiation specificity and it presents the characteristics of the above-described marker. Furthermore, this enzyme is quite stable and persists for a long time in samples, even at ambient temperatures. Therefore, sampling may be performed without any precautions [5]. Nonetheless, this enzyme presents the disadvantage of being sensitive to enzymatic inductors, which may hinder the study of promoters which are often enzymatic inductors.

But different isoenzymes may be interesting for clinical investigation. In particular, $GGT_1$ of prealbumin mobility is frequent in primary liver cancer in man [6]. So GGT increase in rat carcinogenesis seems to be an isoenzyme marker of dedifferentiation, different from the isoenzyme increase caused by inductors. In the study of promoters, which are enzymatic inductors such as phenobarbital, increased GGT has both origins. In the rat, GGT activity in the liver homogenate is slightly increased by phenobarbital alone [7]. Such a large increase in tumor promotion may be attributed to dedifferentiation.

### *Study of a Particular Case*

Studies using induction of carcinogenic lesions in the rat liver showed the perfect correlation of plasmatic GGT with these lesions. One of them is reported here.

#### *Protocol*

The first study was performed according to the following protocol.

120 Fisher 344 male rats weighing about 170 g were allocated to eight different groups of 10 or 20 animals each. One group was untreated and served as control, whilst the other groups received N2-fluorenylacetamide (2-FAA) as the initiating agent in food at 0.02%. Groups 1, 4 and 5 were treated with 2-FAA for 8, 12 and 20 weeks, respectively. Groups 2 and 3 were submitted to recovery periods of 4 and 12 weeks after receiving 2-FAA for 8 consecutive weeks. Groups 6 and 7 were treated with phenobarbital (PhB) at 50 mg/kg by daily gavage for 4 and 12 weeks, respectively, after 8 weeks of 2-FAA treatment.

Before being killed by exsanguination at the abdominal aorta under sodium pentobarbitone anesthesia, each rat received six injections of an iron colloidal solution (100 mg/ml as 0.125 ml s.c. for 100 g of live weight) every 2 days. The aim was to induce hepatic siderosis and reveal lesions not storing iron with the Prussian blue reaction. Samples of blood and liver were taken at autopsy, on which study of GGT was performed.

Lesions of the liver were noticed on the four main lobes: left, medial left, medial right and right anterior.

#### *Anatomopathological Results (table I)*

At 8 weeks of 2-FAA feeding, the number of foci was 57.4 and there were 1.7 neoplastic nodules per rat liver. Further administration of 2-FAA led to an increase in the number of foci from week 8 to 12, and from week 12 to

*Table I.* Results of promotion-initiation study (mean and SEM)

| Group | Anatomopathological (total per liver) | | Biochemical: GGT | |
|---|---|---|---|---|
| | foci | nodules | plasma, IU/l | liver, IU/kg |
| C | 0 (0) | 0 (0) | 1 (1) | 96 ( 11.1) |
| 1 | 57.4 (6.71) | 1.7 (0.32) | 5 (0.6) | 2,676 (157.5) |
| 2 | 5.9 (1.02) | 0.7 (0.2) | 2 (0.3) | 887 ( 93) |
| 3 | 6.1 (1.29) | 0.2 (0.14) | 3 (0.5) | 1,265 (119.5) |
| 4 | 81.1 (5.95) | 1.8 (0.36) | 27 (7.1) | 5,383 (557.1) |
| 5 | 72 (7.94) | 2.4 (0.54) | 50 (6.1) | 7,366 (391.1) |
| 6 | 17.8 (1.54) | 1.9 (0.59) | 8 (1.5) | 2,560 (294.4) |
| 7 | 14.7 (1.85) | 1.6 (0.45) | 10 (9) | 3,890 (339.5) |

*Table II.* Correlations between GGT, foci and nodules

| | Foci | Nodules | GGT of plasma | GGT of liver |
|---|---|---|---|---|
| Foci | 1 | – | – | – |
| Nodules | 0.43* | 1 | – | – |
| GGT of plasma | 0.57* | 0.38* | 1 | – |
| GGT of liver | 0.67* | 0.48* | 0.87* | 1 |

* $p < 0.01$.

20 (but there were no statistical differences at weeks 12 and 20 in comparison with week 8). With both recovery and phenobarbital administration, after 8 weeks of 2-FAA, the number of foci decreased sharply between weeks 8 and 12 and then abated between weeks 12 and 20. Regression of foci was faster with recovery than with phenobarbital; both regression curves showed marked individual variations. With phenobarbital, the number of nodules remained at the level reached after 8 weeks of 2-FAA; in contrast, this number decreased sharply during the first 4 weeks of recovery, and then continued to decrease between 4 and 12 weeks.

*Biochemical Results (table I)*

*With 2-FAA.* GGT showed a strong increase in the liver and plasma with a time-related effect. This treatment also increased alkaline phosphatases in the liver and slightly in the plasma. This parameter appeared as an initiation marker when compared with the recovery of phenobarbital after 2-FAA initiation, which brought it back to a basal dose level and therefore it did not appear as a good marker for the past initiating effect. During the first 12 weeks, 2-FAA alone slightly increased total protein in the liver. It decreased it in the plasma during the same period. Transaminase and LDH of liver and plasma failed to reveal any variation induced by 2-FAA.

*With Recovery.* After initiation with 2-FAA, the level of hepatic and plasmatic GGT and alkaline phosphatases regressed, particularly after 4 weeks, and remained stationary between 4 and 12 weeks. GGT regressed only partially and so appeared as a marker of past initiation; alkaline phosphatases returned to basal levels. Variations of total protein were moderate and not significant.

*With Phenobarbital (PhB).* The levels of hepatic and plasmatic GGT increased with a time-related effect and were in the intermediate position

*Table III.* Correlations

| | Foci | Nodules | Carcinoma |
|---|---|---|---|
| *With all animals killed at 37 weeks (n = 88)* | | | |
| GGT plasma | 0.30* | 0.29* | 0.70* |
| *Without animals presenting carcinoma (n = 43)* | | | |
| GGT plasma | 0.50* | 0.33** | |

* $p < 0.01$.
** $p < 0.05$.

between the levels of 2-FAA and those of recovery at 12 and 20 weeks after the beginning of the experiment, with statistically significant inter-group differences each time. These variations of GGT were time-related with 2-FAA and with both recovery and PhB. Furthermore, the correlations established with individual data between GGT and liver lesions (foci and nodules) were also studied and were very significant (table II).

## *Conclusion*

Plasma levels of GGT could be a good estimation of foci and nodules of liver and could be a good parameter for limited in vivo bioassays of carcinogenesis in rats. This parameter could allow a more complete study with regard to the kinetics of apparition of lesions and an earlier response by biochemistry, which is a more rapid and easier method than pathology. Moreover, the kinetic variation of GGT could modulate the period of necropsy in bioassays.

This parameter was used in a limited in vivo carcinogenesis bioassay of some benzodiazepines [3, 4] which failed to reveal any inhibiting or promoting effect of these compounds. The promoting effect of DDT and PCB alone or in association with phenobarbital after FAA feeding was also tested in a similar study and confirmed the good correlation between plasmatic GGT and pathology results (table III). Correlations between anatomopathological and biochemical results after 37 weeks seem to be as good as in the above experiment, with, in addition, the problem of evaluating the relative importance of the three types of lesions in such a statistical study. GGT is highly correlated with carcinomas and less, but also well correlated, with foci and nodules. This proves that GGT is a carcinogen marker in the rat and that enzymatic induction is not of much importance compared with carcinogenesis. Therefore, this study confirmed the interest presented by GGT, a carcinogen marker, and showed that this parameter may also be used in order to quantify the process in rats.

## *References*


1 Braun, J.P.; Rico, A.G.; Bernard, P.; Burgat Sacaze, V.: La gamma glutamyl transférase: de la pathologie humaine à la biologie animale. Annls Biol. clin. *359:* 433–457 (1977).

2 Ketterer, B.: Biochemical markers in carcinogenesis; in Gorrod, Testing for toxicity, pp. 351–365 (Taylor & Francis, London 1981).

3 Mazué, G.; Remandet, B.; Gouy, D.; Berthe, J.; Roncucci, R.; Williams, G.A.: Limited in vivo bioassays of some benzodiazepines: lack of experimental initiating or promoting effect of the benzodiazepine tranquillizers diazepam, clorazepate, oxazepam and lorazepam. Archs int. Pharmacodyn. Thér. *257:* 59–65 (1982).

4 Remandet, B.: Les benzodiazepines sont-elles carcinogènes; thèse Toulouse (1981).

5 Rosalki, S.B.; Nemesansky, E.; Foo, A.Y.: A new fluorescence method for gamma-glutamyl transferase isoenzyme demonstration. Ann. clin. Biochem. *18:* 25–27 (1981).

6 Siest, G.; Heusghem, C.: Gamma glutamyl transferases. Adv. Biochem. Pharmacol. (Masson, Paris 1982).

7 Stewart, H.L.; Williams, G.M.; Keyser, C.H.; Lombard, L.S.; Montali, R.J.: Histologic typing of liver tumors of the rat. J. natn. Cancer Inst. *64:* 177–207 (1980).

8 Williams, G.M.; Klaiber, M.; Parker, S.E.; Farber, E.: Nature of early appearing, carcinogen-induced liver lesions resistant to iron accumulation. J. natn. Cancer Inst. *57:* 157–165 (1976).

9 Williams, G.M.; Yamamoto, R.S.: Absence of stainable iron from preneoplastic and neoplastic lesions in rat liver with 8-hydroxyquinoline-induced siderosis. J. natn. Cancer Inst. *49:* 685–692 (1972).


Dr. Guy Mazué, Centre de Recherches Clin Midy, rue du Professeur-Blayac,
F-34082 Montpellier Cédex (France)

Safety Evaluation and Regulation of Chemicals. 1st Int. Conf., Boston, Mass., 1982, pp. 259–269 (Karger, Basel 1983)

# Evaluation of an in vitro Cell Toxicity Test Using Rabbit Corneal Cells to Predict the Eye Irritation Potential of Surfactants

*Helen North-Root, Frank Yackovich, Janis Demetrulias, Maximo Gacula, Jr., John E. Heinze*

Armour-Dial Company, Armour Research Center, Scottsdale, Ariz., USA

The Draize rabbit eye irritation test [7] has been the primary method used to assess the potential eye irritation of most chemicals and consumer products. Over the years, this method has received much criticism due to lack of interlaboratory consistency and limited correlation with human experience; however, it is considered useful for screening potentially severe eye irritants. Unfortunately, the testing of chemicals may cause pain to the animal, as well as severe and, in some instances, permanent damage to the eye. For these reasons, considerable interest has been generated and substantial funds have been committed by industry toward identifying alternative methods for screening potential eye irritants without using live animals. This paper describes a simple in vitro method for screening surfactants for cell toxicity in rabbit corneal cells maintained in continuous culture. The results obtained with 13 surfactants indicate a strong correlation between relative in vitro cell toxicity and relative in vivo eye irritation measured in the rabbit. Thus, the cell toxicity test using rabbit corneal cells appears to have the potential to identify severely irritating compounds without using live animals.

Surfactants were selected for testing since they are widely used in cosmetic preparations for their detergent, emulsifying, germicidal, wetting and foaming characteristics. In addition, a considerable amount of information exists on the ability of various surfactants to cause irritation to the rabbit eye [6, 12, 14]. For example, *Hazleton* [14] evaluated a wide variety of surfactants for their ability to produce eye irritation and reported the eye irritation potential of surfactants to follow the general order of cationic > anionic > nonionic. He stated that the ionic characteristics, rather than pH, reduction in surface tension or wetting power appeared to be the best initial criteria for predicting eye irritation.

A number of reports in the literature concern the various effects of surfactants on in vitro systems. For example, *Choman* [4] investigated the ability of surfactants to increase swelling in discs cut from calf skin. His hypothesis was that swelling would be directly proportional to irritation; however, the anionic surfactants were the only materials which had an effect. *Bettley* [2] attempted to correlate the effects of various surfactants on potassium permeability in yeast, hemolysis and potassium release in erythrocytes, and morphology in buccal mucosa cells. The ability of the surfactants to increase potassium permeability and to cause potassium release and hemolysis again followed the general order of cationic > anionic > nonionic. *Prottey and Ferguson* [16] reported that histamine release by rat peritoneal mast cells in vitro occurred in response to both anionic and nonionic surfactants. These investigators also studied the effects of surfactants upon DNA and triglyceride synthesis and release of cytoplasmic proteins by guinea pig kidney fibroblasts [9]. They found that surfactants which were considered more irritating caused larger increases in DNA and triglyceride synthesis.

*Gibson* [11] evaluated the ability of several surfactants to cause lysis of rabbit polymorphonuclear leukocyte granules in vitro in relation to their ability to produce skin irritation in the rat. He found that the nonionic surfactants caused minimal to no granule lysis and minimal skin irritation in the rat, while the anionic surfactants caused marked granule lysis and considerable skin irritation in vivo. The correlation between in vitro and in vivo activity was not as good for two of the cationic surfactants which were studied. *Ernst and Arditti* [8] examined a homologous series of nonionic surfactants in HeLa cells and found that toxicity increased with an increase in chain length of 10 to 18 carbon atoms. Lethal levels of nonionic and amphoteric surfactants coincided with a reduction in the media surface tension to 45 $dyn \cdot cm^{-1}$ or less, suggesting that the physical properties of the surfactants were a primary cause of their toxic effects on HeLa cells.

The results of these investigations indicated that several in vitro processes, including cell toxicity, were representative of the interaction of surfactants with skin. More recently attention has focused on in vitro procedures for screening eye irritants. *Bell* et al. [1] used the human buccal mucosa system [2] to examine the loss of nuclei in response to treatment with various alkyl sulfates. Other surfactant types were not examined. Moreover, the relevance of this system to eye irritation is not clear. *Burton* et al. [3] utilized the enucleated rabbit eye placed in a saline-perfused temperature-controlled chamber. The only surfactant they studied, cetyl trimethylammonium bromide, which is reportedly irritating to rabbit eyes [5], was classified as a 'slight irritant'. In addition, the system did not eliminate the involvement of live animals as eye donors.

In developing the in vitro system described here, cell toxicity was chosen as the means for measuring surfactant effects since it is a conveniently measurable property. Cell toxicity should also be a relevant parameter since compounds which are highly toxic to cells in vitro are likely to be highly irritating to the eye in vivo.

To increase the ease and reproducibility of measuring cell toxicity, an established cell line was chosen. The use of an established line eliminates the need to obtain cells continuously from live animals for primary cultures. Since the objective of developing an in vitro system was to evaluate eye irritation potential, a cell line derived from ocular tissue was thought to be most appropriate. The SIRC line, derived from the corneal cells of rabbits, was available through the American Type Culture Collection (Rockville, Md.). The rabbit corneal cell in culture would seem to be the most representative in vitro analog of the cornea, a major structural component of the eye which also receives the most weighting in the Draize scoring system due to its importance to clear vision. For these reasons, the SIRC rabbit corneal cell line was chosen as the system in which to measure cell toxicity.

In order to rely upon results obtained in an in vitro system, the in vitro results must be quantitatively validated against 'real life' results. While human data on eye irritation would be the most relevant for this purpose, well-documented reports on human exposure to surfactants are rare. Although published data on rabbit eye irritation produced by surfactants are more widely available, insufficient data were available to compare eye irritation scores directly to cell toxicity scores. For instance, while *Hazleton* [14] and *Draize and Kelly* [6] ranked various surfactants according to irritation potential or maximum tolerated concentration, respectively, no Draize scores were reported. In the study described here, eleven surfactants from the three major ionic groups, as well as two amphoteric compounds, were screened for toxicity in the rabbit corneal cell line and for eye irritation in the New Zealand albino rabbit.

## *Materials and Methods*

### *Cationics*

Alkyldimethylbenzyl ammonium chloride (Roccal) was obtained from Hilton-Davis Chemical Co. (Cincinnati, Ohio), methyldodecylbenzyltrimethyl ammonium chloride (MAC) was obtained from Chem Service (Media, Pa), 1-[(2-hydroxyethyl) carbamoyl] methyl pyridinium chloride stearate (Emcol E607S) and 1-[(2-hydroxyethyl) carbamoyl] methyl pyridinium chloride laurate (Emcol E607L) were obtained from Witco Chemical Corp. (New York, N.Y.), and diisobutylphenoxyethoxyethyldimethylbenzyl ammonium chloride NF (Hyamine 1622) was obtained from Rohm and Haas (Philadelphia, Pa.).

*Anionics*

Sodium lauryl sulfate (Richonol A) and triethanolamine lauryl sulfate (Richonol T) were obtained from Richardson Co. (Des Plaines, Ill.) and sodium $C_{14-16}$ olefin sulfonate (Bioterge AS 40) was obtained from Stepan Chemical Co. (Northfield, Ill.)

*Amphoterics*

Cocoamidopropyl dimethyl glycine (Emery 6748A) was obtained from Emery Industries, Inc. (Cincinnati, Ohio) and cocoamphocarboxy propionate (Miranol C2M SF conc.) was obtained from Miranol Chemical Co., Inc. (Irvington, N.J.).

*Nonionics*

Polyoxyethylene (23) lauryl ether (Brij 35), polyoxyethylene (20) sorbitan monopalmitate (Tween 40) and polyoxyethylene (20) sorbitan monostearate (Tween 60) were obtained from ICI Americas (Wilmington, Del.).

All surfactant concentrations or other values reported here are expressed in terms of percent active ingredient.

*Cells and Cell Culture*

The SIRC cell line was derived from the cornea of a normal rabbit *(Orycytologus cuniculus)* by *M. Volkert* of the Staatens Seruminstitut, Copenhagen, Denmark, in 1957. The cell line has since undergone over 400 passages and was obtained from the American Type Culture Collection (CCL 60). Cell stocks are stored in liquid nitrogen. The cells were grown in Ham's F12 medium (KC Biological, Inc., Lenexa, Kans.) supplemented with 10% heat-treated (30 min, 56°C) donor calf serum (Flow Laboratories, Inc., Inglewood, Calif.) at 37°C in a humidified atmosphere containing 5% $CO_2$ in air. The medium was adjusted to pH 7.0 prior to filter sterilization.

*Preparation of Cells*

The medium was removed from the 100-mm Petri dishes containing the growing cells. The tissue culture dish was washed with 1 ml of saline $D_1$ [15] containing 0.05% trypsin (1:300; Gibco Laboratories, Santa Clara, Calif.) and 0.02% disodium ethylenediamine tetraacetic acid (EDTA); 2 ml of the trypsin-EDTA solution were added to the dish which was then incubated for 20 min in the 37°C humidified incubator containing 5% $CO_2$ in air. The cell clumps were broken with a Pasteur pipette. 2 ml Ham's F12 medium supplemented with 10% donor calf serum were added to the dish to stop the action of trypsin, and the cells were thoroughly mixed with a Pasteur pipette. Cell numbers were determined with an Elzone particle counter (Particle Data, Inc., Elmhurst, Ill.).

For passage of the cell culture, $10^4$ cells were added to 100-mm dishes containing 14 ml Ham's F12 medium supplemented with 10% heat-treated donor calf serum. The medium was changed twice a week.

*Preparation of Surfactant Solutions*

Emcol E607L, Emcol E607S and MAC were dissolved in sterile glass-distilled water on a weight/volume basis and allowed to 'self-sterilize' for at least 1 h. The pH was not adjusted. Further dilutions were made in sterile Ham's F12 medium without serum on a volume/volume basis. The remaining surfactants were dissolved directly in Ham's F12 medium without serum (w/v) and the pH was adjusted to 7.0. Tween 60 was autoclaved at 121 °C for 15 min and the rest of the surfactant solutions were filter-sterilized. Further dilutions were made with sterile Ham's F12 medium without serum (v/v).

*Treatment of Cells*

The cells were adjusted to $1 \times 10^4$ cells/ml in Ham's F12 medium containing no serum to make pipetting easier. 40 μl of the cell suspension, i.e. 400 cells, were added to 60-mm Petri dishes containing 5 ml Ham's F12 medium supplemented with 10% heat-treated donor calf serum. The dishes were incubated at 37 °C in a humidified incubator containing 5% $CO_2$ in air for 18–20 h to allow the cells to attach to the dish. The medium was removed and the dishes were washed twice with 2 ml Ham's medium without serum. 5 ml of each concentration of surfactant in Ham's F12 medium without serum were added to triplicate plates. Six controls containing Ham's F12 medium without serum were included in each run. A minimum of ten concentrations was tested in each run. The dishes were incubated at 37 °C in a humidified atmosphere containing 5% $CO_2$ in air for 1 h. The medium was removed from all dishes which were then washed twice with 2 ml Ham's F12 medium supplemented with 10% heat-treated donor calf serum. 5 ml of Ham's F12 medium with 10% heat-treated donor calf serum were then added to each dish. The dishes were placed in the 37 °C humidified incubator containing 5% $CO_2$ in air for 7–8 days. The medium was removed from the dishes. The cells were then fixed with 10% formalin and stained with 0.1% crystal violet. Colonies containing 10 or more cells were counted.

*Animals*

New Zealand albino rabbits *(Orycytologus cuniculus)* of both sexes were obtained from 'A' Rabbitry, Tucson, Ariz., and housed individually in self-flushing Hoeltge wire mesh cages. They received 12 h of light/12 h of darkness daily and Purina rabbit chow (Ralston Purina, St. Louis, Mo.) and tap water ad libitum. The rabbits were identified with metal ear tags and were acclimated to the laboratory for a minimum of 1 week before assignment to experimental groups. The animals were assigned randomly to treatment groups and 3 animals of each sex comprised a treatment group.

*Test Material Administration*

All the surfactants were prepared as aqueous solutions on a weight/volume basis and dilutions of each stock solution were prepared on a volume/volume basis. To minimize the potential occurrence of severe eye irritation, the cationic surfactants were tested at concentrations no higher than 6%, and the anionic, amphoteric and nonionic surfactants were tested at concentrations not exceeding 30%. All solutions were prepared on the day of dosing. 0.01 ml of the appropriate surfactant concentration was applied directly to the cornea of each rabbit [13] with a 0.25-$cm^3$ glass syringe equipped with a 20-gauge ball-tipped needle. Each rabbit was then fitted with a flexible plastic Elizabethan-type collar which remained in place until the first observation period 24 h later.

*Table I.* In vitro $LD_{50}$ values and surfactant concentrations predicted to cause Draize scores of 20

| Surfactants | $LD_{50}$ values ppm | Surfactant concentration predicted for Draize score = 20 % |
|---|---|---|
| Cationic | | |
| Roccal | 2.2 | 2.5 |
| MAC | 2.5 | 3.5 |
| Emcol E607S | 2.6 | 28.8 |
| Emcol E607L | 11.8 | 6.1 |
| Hyamine 1622 | 4.1 | 3.5 |
| Anionic | | |
| Richonol A | 8.5 | 25.1 |
| Richonol T | 11.8 | 29.0 |
| Bioterge AS 40 | 12.4 | 33.8 |
| Amphoteric | | |
| Emery 6748A | 8.5 | 18.7 |
| Miranol C2M SF | 35.5 | ~ 90.0* |
| Nonionic | | |
| Brij 35 | 36.0 | >>> 90.0* |
| Tween 40 | 11,500.0 | > 90.0* |
| Tween 60 | 36,000.0 | >> 90.0* |

* (~ 90.0) < (> 90.0) < (>> 90.0) < (>>> 90.0).

*Grading Eye Irritation*

Corneal, iris and conjunctival responses were evaluated using the scale described by *Draize* et al. [7]. Scoring was done at 24, 48 and 72 h following application of each surfactant. Fluorescein stain was used to aid in determining the extent of corneal damage.

*Statistical Analysis*

For the in vitro data, the $LD_{50}$ value, i.e. the surfactant concentration allowing 50% cell survival, was determined using the probit method [10]. The proportion of cells responding (killed) was adjusted for the natural response rate obtained from two control samples of three plates each. The adjustment procedure is described by *Finney* [10]. The estimate of the $LD_{50}$ was obtained by first computing $\hat{x} = (5.0 - Y)/b$, where Y is the intercept of a regression equation and b is the regression coefficient. Then, the $LD_{50}$ = exponential of 2.3025 $\hat{x}$.

For the in vivo data, the mean Draize score, calculated from individual scores of 6 rabbits, was plotted vs. each surfactant concentration tested to obtain a dose-response plot. A Draize score equal to 20 units from a possible total of 110 was selected as a representative score within the range of scores actually observed during evaluation of the majority of the surfactants. The

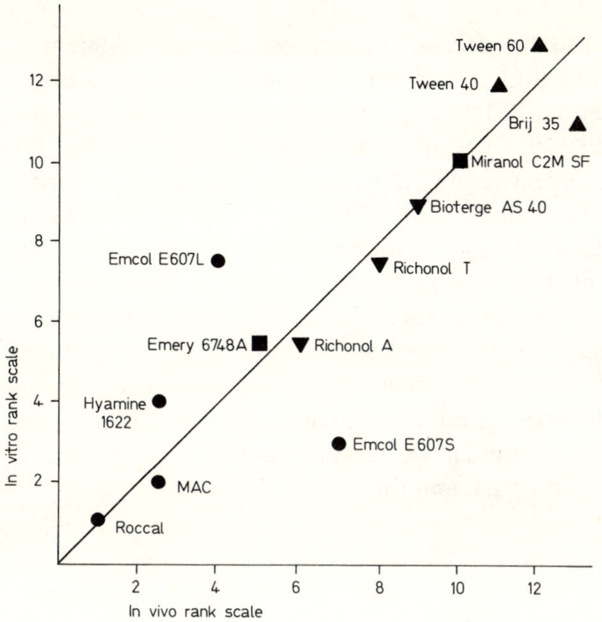


*Fig. 1.* Toxicity/irritation ranking of surfactants. Thirteen surfactants were ranked from 1 (most toxic/irritating) to 13 (least toxic/irritating). Rank correlation analysis indicated a correlation coefficient of 0.90 (significant at $p < 0.01$). ● = Cationic surfactants; ■ = amphoteric surfactants; ▼ = anionic surfactants; ▲ = nonionic surfactants.

surfactant concentration predicted to cause a Draize score of 20 was then used to rank the surfactants from 1 (most irritating) to 13 (least irritating). Likewise, the $LD_{50}$ values were assigned ranks with the lowest $LD_{50}$ given a rank of 1 (most toxic) and the largest $LD_{50}$ given a rank of 13 (least toxic). The ranks for the predicted surfactant concentration which would cause a Draize score of 20 and the $LD_{50}$ values were then correlated using the Spearman rank correlation analysis as described in *Steel and Torrie* [17].

## *Results*

Table I shows the $LD_{50}$ value obtained for each of 13 surfactants in the rabbit corneal cells and the concentration of each surfactant predicted to cause a Draize score of 20 in the rabbit eye in vivo. The estimates of the in vivo surfactant concentrations for Miranol C2M SF, Tween 40, Brij 35 and

Tween 60 were based on one concentration level using the direct corneal application of 0.01 ml and therefore should be considered approximations. For these chemicals, a straight line was drawn through the origin to obtain the predicted concentration. Brij 35 and Tween 60 were also tested undiluted (active concentration of 97%) using the established Draize method of instilling 0.1 g or 0.1 ml, respectively, into the conjunctival sac of 6 rabbits. The resulting Draize scores of 10.2 for Brij 35 and 9.0 for Tween 60 indicated that these materials were, in fact, nonirritating and they were ranked accordingly.

Using the data in table I, a rank correlation analysis was performed and the results are depicted graphically in figure 1. The correlation coefficient was found to be 0.90 which was significant at $p < 0.01$ ($t_{11DF} = 6.74$). With a few exceptions, the order of eye irritation, as well as cytotoxicity, was cationic > anionic = amphoteric > nonionic. The two surfactants showing the least correlation between in vivo and in vitro results were the two pyridinium chloride compounds, Emcol E607L and Emcol E607S.

*Discussion*

A straightforward and readily handled in vitro system has been developed for predicting relative eye irritation by measuring cell toxicity in SIRC rabbit corneal cells. Use of this method can reduce the amount of eye irritation testing performed in live animals without increasing the cost or man-hours required to obtain results. For example, to evaluate one surfactant, approximately 12 man-hours and $40 in material costs were required when the SIRC cell system was used vs. 10 man-hours and $256 in material costs when the modified Draize procedure was used.

A significant correlation between the relative toxicity in rabbit corneal cells in vitro and the relative eye irritation in a modified Draize test was seen in response to treatment with 13 surfactants. These surfactants included representatives of the cationic, anionic, nonionic and amphoteric groups. Moreover, the in vivo results are consistent with published data on the relative eye irritation potential of surfactants in rabbits [6, 12, 14] and with more limited human eye irritation data [12].

The degree of correlation between the in vitro and in vivo systems described here is excellent despite the wide variety of surfactants which was evaluated. Only two surfactants, Emcol E607L and Emcol E607S, the two closely related cationic pyridinium chloride materials, did not behave as predicted. Emcol E607L was less toxic to rabbit corneal cells than expected from the eye irritation results. A possible explanation for the discrepancy which was investigated was the different pH values of the surfactant solutions used in the two tests. For the SIRC cell test, surfactants were diluted in tissue culture medium to a final pH of 7.2; in the eye irritation test, surfactants were diluted in water and the resulting pH for Emcol E607L was 2.5. However, the pH of the Emcol E607L solution used in the rabbit eye experiments did not appear to have an effect on the resultant irritation, i.e., adjusting the pH from 2.5 to 6.5 did not change the Draize scores significantly and substantial eye irritation was still noted. Emcol E607S, in contrast, was relatively more toxic to rabbit corneal cells than expected from the rabbit eye irritation scores. Emcol E607S is relatively insoluble in aqueous systems. This property may have contributed to the relatively high in vitro cell toxicity of Emcol E607S by making it difficult to completely rinse the material off the SIRC cells after the standard 1-hour exposure period. Finally, the greater molecular weight of Emcol E607S may have resulted in less penetration of the material into the rabbit eye with correspondingly lower Draize scores than were achieved with Emcol E607L.

The major limitations of using an in vitro screening method such as the rabbit corneal cell system are the lack of a functioning organ complete with structural components and the inability to examine reversibility of damage. The SIRC cells have no vasculature; therefore, any effects of chemicals on conjunctival membranes or the iris cannot be measured. However, the method used in this study for surfactant administration, i.e. direct corneal application of a small volume of test material [13], may limit damage primarily to the cornea. Moreover, the Draize scoring system weights corneal damage the heaviest. In fact, when corneal scores alone were used to rank six of the surfactants in the in vivo portion of this study, the same relative rankings were achieved as when the complete Draize scores were used.

Reversibility of damage cannot be assessed if the measurement of toxicity is limited to cell death (or lysis). However, there are other parameters which can be examined in vitro which do not affect cell viability and which, with time, might indicate reversibility to the original, 'normal', state. Examples of such parameters include changes in cell morphology [2], alterations in intracellular biochemical processes [9], changes in rate of cell division, and cellular uptake or release of specific biochemical indicators [2, 16]. These and other parameters could be evaluated readily in the rabbit corneal cell system.

In conclusion, an in vitro system which accurately predicts the relative eye irritation potential of surfactants has been described. Further evaluation of the SIRC rabbit corneal cell system is necessary to determine its suitability other laboratories would be helpful to determine more fully interlaboratory reproducibility. However, no in vitro procedure is likely to supplant completely an in vivo system which takes into consideration the wide range of factors which can influence ocular toxicity. The close agreement between the in vitro and in vivo results presented here supports the contention that an appropriate in vitro method can be valuable in the preliminary stages of safety evaluation and further, that it can be useful in the selection of product formulations for in vivo testing.

## *References*


1 Bell, M.; Holmes, P.M.; Nisbet, T.M.; Uttley, M.; Van Abbe, N.J.: Evaluating the potential eye irritancy of shampoos. Int. J. cosmet. Sci. *1:* 123–131 (1979).

2 Bettley, F.R.: The toxicity of soaps and detergents. Br. J. Derm. *80:* 635–642 (1968).

3 Burton, A.B.G.; York, M.; Lawrence, R.S.: The in vitro assessment of severe eye irritants. Fd Cosmet. Toxicol. *19:* 471–480 (1981).

4 Choman, B.R.: Determination of the response of skin to chemical agents by an in vitro procedure. II. Effects of aqueous anionic, cationic, and nonionic surfactant solutions. J. invest. Derm. *40:* 177–182 (1963).

5 Dewey, W.L.; Malone, M.E.; Harris, L.S.; Bottom, I.: Pharmacologic and toxicologic evaluation of 2-(N-tetradecyl-N-ethylamino)-2-methyl-1, 3-propanediol hydrochloride, an inhibitor of intestinal absorption. Fd Cosmet. Toxicol. *12:* 115–131 (1968).

6 Draize, J.H.; Kelly, E.A.: Toxicity to eye mucosa of certain cosmetic preparations containing surface-active agents. Proc. scient. Sect. Toilet Goods Ass. *17:* 1–4 (1952).

7 Draize, J.H.; Woodard, G.; Calvery, H.O.: Methods for the study of irritation and toxicity of substances applied topically to the skin and mucous membranes. J. Pharmac. exp. Ther. *82:* 377–390 (1944).

8 Ernst, R.; Arditti, J.: Biological effects of surfactants. IV. Effects of non-ionics and amphoterics on HeLa cells. Toxicology *15:* 233–242 (1980).

9 Ferguson, T.F.M.; Prottey, C.: The effects of surfactants upon mammalian cells in vitro. Fd Cosmet. Toxicol. *14:* 431–434 (1976).

10 Finney, D.J.: Probit analysis; 3rd ed., p. 125 (Cambridge University Press, Cambridge 1971).

11 Gibson, W.T.: Lysis of rabbit polymorphonuclear leucocyte granules by surfactants of differing structure and irritancy. Fd Cosmet. Toxicol. *18:* 511–515 (1980).

12 Grant, W.M.: Toxicology of the eye; 2nd ed., pp. 962–967 (Thomas, Springfield 1974).

13 Griffith, J.F.; Nixon, G.A.; Bruce, R.D.; Reer, P.J.; Bannan, E.A.: Dose-response studies with chemical irritants in the albino rabbit eye as a basis for selecting optimum testing conditions for predicting hazard to the human eye. Toxicol. appl. Pharmacol. *55:* 501–513 (1980).

14 Hazleton, L.W.: Relation of surface active properties to irritation of the rabbit eye. Proc. scient. Sect. Toilet Goods Ass. *17:* 5–9 (1952).
15 O'Neil, J.P.; Hsie, A.W.: Phenotypic expression time of mutagen induced 6-thioguanine resistance in Chinese hamster ovary cells. Mutat. Res. *59:* 109–118 (1979).
16 Prottey, C.; Ferguson, T.F.M.: The effects of surfactants upon rat peritoneal mast cells in vitro. Fd Cosmet. Toxicol. *14:* 425–430 (1976).
17 Steel, R.G.D.; Torrie, J.H.: Principles and procedures of statistics, p. 409 (McGraw-Hill, New York 1960).

Dr. Helen North-Root, Armour-Dial Company, Armour Research Center,
15101 N. Scottsdale Road, Scottsdale, AZ 85260 (USA)

Safety Evaluation and Regulation of Chemicals. 1st Int. Conf., Boston, Mass., 1982, pp. 270–276 (Karger, Basel 1983)

# Investigation of *Drosophila* for Teratogenic Screening

*Ronald L. Schuler, Bryan D. Hardin, Richard W. Niemeier*[1]

Department of Health and Human Services, US Public Health Service, Centers for Disease Control, National Institute for Occupational Safety and Health, Division of Biomedical and Behavioral Science, Robert A. Taft Laboratories, Cincinnati, Ohio, USA

The complex mating procedures, scheduling of treatments, requirements for trained personnel and resultant costs of traditional mammalian teratological tests make it difficult, if not impossible, to study completely the large number of new chemicals that appear in commerce. The backlog for testing of existing chemicals is already overwhelming. Because of this, a number of investigators in recent years have sought to develop rapid test systems that attempt to screen large numbers of chemicals in order to identify the more hazardous substances. Priority chemicals are then designated for more rigorous and definitive examination.

*Drosophila* has the potential to provide such a screen. *Wilson* [8] states that *Drosophila* is particularly noteworthy, in that it is capable of absorbing, circulating, metabolizing, and excreting chemicals while economically producing large numbers of offspring in a relatively short period of time.

Previously, *Drosophila* had been studied for morphological effects on adults following larval treatment with vinblastine (VB) [2], 5-bromodeoxyuridine (BDU) [3], and thymidine (TD) [6]. VB induced primarily eye deformities and reduction of antennae, wings, halteres and the thorax. BDU caused deformities of the mouthparts and rear legs; wings were reduced. TD induced wing notches or cuts, extra bristles, abdominal irregularities, and leg malformations. In our laboratory, a standardized test approach has been developed to use *Drosophila* as a preliminary screening phase in teratological investigation [7]. The method is simple, having clearly defined scoring

[1] The authors wish to thank Ms. *Mary Radike* for her technical assistance throughout this work, and Ms. *Sandra Clark* for her aid in preparing this manuscript.

parameters. Data generated from treated groups are compared with control group data using the chi-square statistical test.

## *Materials and Methods*

Adult flies mate and females deposit eggs on the surface of an instant *Drosophila* medium into which the test chemical has been incorporated. Within 24 h the eggs hatch and the larvae immediately begin to feed on the medium. Following metamorphosis through three larval stages and one pupal stage, adult flies emerge from their pupal cases in about 10 days. These flies are examined under a binocular microscope for morphological abnormalities. Incidences of abnormalities found in treated flies are compared to those of concurrent control flies using the chi-square test for significance. Details of materials and methods used can be found in an earlier paper [5].

## *Results*

The *Drosophila* system has yielded a set of distinct abnormalities arising from treatment with a number of different test chemicals. Table I is a partial list of the chemicals investigated together with some of the more significant anomalies found. Included in this table is the amount of chemical added to each treatment vial. It must be emphasized that these are rough estimates generated during developmental work and should not be used for statistical comparisons. Table II lists the first three attempts to validate this system using 6-aminonicotinamide, cyclophosphamide and hydroxyurea (all known teratogens). Two tests were negative, i.e., differences between test and control data were not significant using the chi-square test ($p < 0.05$), the exception being hydroxyurea which showed a high number of reduced bristles. Of the more striking responses seen, bent bristles are particularly noteworthy in N,N-dimethylacetamide-treated flies (40.4% incidence) and in dimethylsulfoxide-treated flies (up to 70% incidence); control incidences are less than 1%. Wing notches, not seen in control flies, were found in 3, 15, 18, and 14% of flies treated with N,N-dimethylacetamide, dimethylsulfoxide, ethylene glycol monoethyl ether, and ethylene glycol monomethyl ether, respectively. Recently, ethylene glycol monoethyl ether [4] and ethylene glycol monomethyl ether [5] have been shown to be teratogenic. These are both ethylene glycol derivatives as are diethylene glycol monobutyl ether and diethylene glycol monoethyl ether (table I) which are currently being investigated by NIOSH for teratogenicity. Table III provides examples of the recurrence of specific malformations in replicate testing. From the data generated

*Table I.* Data derived during developmental work

| Chemical (concentration/g media) | Number of flies examined | Percent of flies with morphologic abnormalities | | | | | | | | | |
|---|---|---|---|---|---|---|---|---|---|---|---|
| | | bristles | | | | irregular shape | | | legs | | wing |
| | | absent | bent | extra | small | head (eye) | thorax | abdomen | absent | twisted | notch |
| Bisphenol A (30 mg/g) | 194 | 1.5 | 3.1 | 0 | 0 | 0 | 3.6 | 6.2 | 0 | 0 | 0.5 |
| Diethylene glycol monobutylether (10 µl/g) | 127 | 3.9 | 1.6 | 0 | 0.8 | 3.9 | 0 | 0.8 | 1.6 | 0 | 0 |
| N, N-dimethyl acetamide (0.25 µl/g) | 183 | 15.3 | 40.4 | 2.7 | 2.2 | 0.5 | 12.0 | 1.6 | 0.5 | 1.6 | 3.3 |
| Ethylene glycol monomethyl ether (7.5 µl/g) | 29 | 3.4 | 3.4 | 3.4 | 0 | 0 | 6.9 | 0 | 0 | 0 | 13.8 |
| Diethylene glycol monoethyl ether (7.5 µl/g) | 221 | 4.5 | 1.8 | 0 | 2.3 | 0.5 | 0.9 | 7.7 | 0 | 0 | 0 |
| Niacin (25 mg/g) | 206 | 0 | 1.5 | 0 | 0.5 | 0 | 1.0 | 8.7 | 0 | 0 | 1.5 |
| Thymidine (3 mg/g) | 82 | 8.5 | 3.7 | 0 | 0 | 0 | 0 | 0 | 0 | 0 | 1.2 |
| Control (water) | 2,183 | 2.8 | 0.5 | – | – | – | – | – | – | – | – |
| | 1,037 | – | – | – | – | – | – | 2.0 | – | – | – |
| | 800 | – | – | 0.4 | 0.5 | 0 | 0 | – | – | – | 0 |

*Table II.* Data derived during validation efforts

| Chemical (concentration/g media) | Number of flies examined | Percent of flies with morphologic abnormalities | | | | | | | | | |
|---|---|---|---|---|---|---|---|---|---|---|---|
| | | bristles | | | | irregular shape | | | legs | | wing |
| | | absent | bent | extra | small | head (eye) | thorax | abdomen | absent | twisted | notch |
| 6-Amino-nicotinamide | | | | | | | | | | | |
| Control | 162 | 4.3 | 0 | 1.2 | 0 | 0 | 0 | 4.3 | 0 | 0 | 0 |
| 5.0 mg/g | 26 | 11.5 | 3.8 | 0 | 0 | 0 | 0 | 0 | 0 | 0 | 0 |
| 4.0 mg/g | 87 | 1.1 | 1.1 | 2.3 | 0 | 0 | 0 | 0 | 0 | 0 | 0 |
| 3.0 mg/g | 82 | 4.9 | 1.2 | 3.7 | 1.2 | 0 | 0 | 6.1 | 0 | 0 | 0 |
| Cyclophosphamide | | | | | | | | | | | |
| Control | 185 | 7.6 | 1.6 | 1.1 | 0 | 0 | 0 | 2.2 | 0 | 0 | 0 |
| 0.10 mg/g | 42 | 19.0 | 0 | 2.4 | 4.8 | 0 | 0 | 0 | 0 | 0 | 0 |
| Hydroxyurea | | | | | | | | | | | |
| Control | 189 | 5.3 | 1.6 | 0 | 0 | 0 | 0 | 0.5 | 0 | 0.5 | 0 |
| 0.2 μl/g | 228 | 6.6 | 5.3 | 2.2 | 7.9 | 0 | 0 | 0 | 0 | 0 | 0 |

*Table III.* Examples of replicate testing results

| Treatment | Number of flies scored | Effects and incidences | | |
|---|---|---|---|---|
| | | humeral bristle defects | wing notches | haltere defects |
| *Dimethysulfoxide* | | | | |
| 5 µl/g media | 79 | 70% | 11% | none |
| 5 µl/g media | 65 | 46% | 20% | none |
| 5 µl/g media | 42 | not scored | 21% | not scored |
| 3 µl/g media | 100 | 13% | 3% | not scored |
| *2-Ethoxyethanol* | | | | |
| 15 µl/g media | 32 | 3% | 19% | none |
| 15 µl/g media | 78 | 3% | 18% | none |
| 15 µl/g media | 21 | 10% | 19% | none |
| 10 µl/g media | 172 | <1% | 10% | none |
| 10 µl/g media | 76 | 5% | 1% | none |
| 10 µl/g media | 35 | 3% | none | none |
| *Sodium heparin* | | | | |
| 1,000 U (solution)/g media | 214 | 34% | not scored | 7% |
| 750 U/g | 38 | 32% | 3% | none |
| 10 mg (solid)/g | 17 | 12% | none | 12% |
| 5 mg/g | 225 | numerous – not counted | none | 5% |
| 5 mg/g | 206 | 62% | none | 8% |
| *Control* | | | | |
| | 187 | 1% | none | none |
| | 237 | none | none | none |
| | 114 | none | none | none |
| | 262 | none | none | none |
| | 342 | <1% | not scored | not scored |
| | 241 | 1% | not scored | not scored |
| | 370 | not scored | none | not scored |
| | 330 | not scored | none | not scored |
| Combined control | – | 0.5% (7/1383) | 0% (0/1500) | 0% (0/424) |

during developmental work, a trend, showing dose-response relationships over several variables or responses, is evident. Specifically, in dimethylsulfoxide-treated flies the incidence of abnormalities increases from 13 to 70% for humeral bristle defects and from 3 to 21% for wing notches when increasing the concentration from 3 to 5 µl/g of medium. Similarly, 2-ethoxyethanol-treated flies exhibited 0–10% wing notch formation at 10 μl/g and 18–19% wing notch formation at a concentration of 15 μl/g of medium. No dose-response relationship with sodium heparin can be derived from these data.

*Discussion*

The majority of data derived from these tests were generated during developmental efforts and therefore complete statistical comparison between treated and control groups is lacking. For the purposes of a teratogen screen, the description of the abnormalities and how they relate to traditional mammalian endpoints is not important. Rather, it is the dramatic and reproducible response of *Drosophila* to chemical treatment during developmental morphogenesis that is important, i.e., a system, essentially new to the field of teratology, has been shown to have potential for such investigations. The utility of this system as a teratogen screen can be evaluated only through the testing of known teratogens and nonteratogens. Such validation efforts are currently underway in this laboratory. In addition, a DNA repair deficient strain (mei-9, mei-41) of *Drosophila* [1] is being studied which may increase the sensitivity of the system.

It is important to note that most of the chemicals produced a specific pattern of abnormalities in treated flies. Although dimethylsulfoxide (DMSO) and ethylene glycol monoethyl ether (EGMEE) both induced wing notch defects at about the same frequency (15 and 18%, respectively), humeral bristle defects were far greater in DMSO-treated flies (59% incidence) than in EGMEE-treated flies (4% incidence). Sodium-heparin-treated flies showed unique abnormalities including humeral knobs (tissue growths near the humeral bristle site), multiple and missing halteres, and the fusion of front leg segments. Such specificity may lend a clue to the as yet unknown mechanism(s) involved.

The methodology of the test system is simple, yet *Drosophila* is well characterized in regard to genetics and embryology thus allowing for further mechanistic studies, e.g., egg-only treatment or individual, larvae-stage testing might be performed. Also, flies having malformations may be mated to investigate the heritability of such effects. Such studies have little relevancy to the development of the system as a screen but this potential may inspire geneticists to pursue *Drosophila* as a tool for further investigation into the cause(s) of the teratological process(es).

There are several items to be considered before adopting this system as a routine screen. Although technical proficiency can be developed within two weeks or so, the degree of objectivity and/or visual skills may vary with the individual. Some dexterity is required for the fine manipulations required while scoring. Scoring in 'blind' is desirable; however, this may be impractical because control vials are, at times, quite distinguishable from treated vials due to reduced numbers and delayed emergence of flies caused from toxic effects at MTD levels. Solubility problems exist; only water-soluble or ethanol-soluble (up to 0.25 ml/vial) chemicals have been successfully administered. Only water-soluble chemicals and controls have been reported here. Other solvents may be used as a vehicle for non-water-soluble substances. But, in view of the malformations induced by dimethylsulfoxide (probably the most widely used solvent for *Drosophila* studies other than water), caution should be exercised in its use and in interpretation of the results.

## *References*

1 Boyd, J.B.; Golino, M.D.; Nguyen, T.D.; Green, M.M.: Isolation and characterization of x-linked mutants of *Drosophila melanogaster* which are sensitive to mutagens. Genetics *84:* 485–506 (1976).

2 Clayton, F.C.; Francœur, R.T.: Some teratogenic effects of vinblastine on the external morphology of *Drosophila melanogaster.* Oncology *25:* 188–192 (1971).

3 Gilbert, E.F.; Pitot, H.C.; Bruyere, H.J.; Cheung, A.L.: Teratogenic effects of 5-bromodeoxyuridine on the external morphology of *Drosophila melanogaster.* Teratology *7:* 205–208 (1973).

4 Hardin, B.D.; Bond, G.P.; Sikov, M.R.; Andrew, F.D.; Beliles, R.P.; Niemeier, R.W.: Testing of selected workplace chemicals for teratogenic potential. Scand. J. envir. Hlth *7:* 66–75 (1981).

5 Nagano, K.; Nakayama, E.; Oobayashi, H.; Yamada, T.; Adachi, H.; Nishizawa, T.; Ozawa, H.; Nakaichi, M.; Okuda, H.; Minami, K.; Yamazaki, K.: Embryotoxic effects of ethylene glycol monomethyl ether in mice. Toxicology *20:* 335–343 (1981).

6 Parkash, O.: On the teratogenic effect of thymidine and its suppression by deoxycytidine in *Drosophila melanogaster.* Experientia *27:* 1089–1090 (1971).

7 Schuler, R.L.; Hardin, B.D.; Niemeier, R.W.: *Drosophila* as a tool for the rapid assessment of chemicals for teratogenicity. Teratol. Carcinog. Mutagen. (in press.).

8 Wilson, J.D.: Review of in vitro systems with potential for use in teratogenicity screening. J. envir. Path. Toxicol. *2:* 149–167 (1978).

Dr. R.L. Schuler, National Institute for Occupational Safety and Health,
4676 Columbia Parkway, Mailstop C-23, Cincinnati, OH 45226 (USA)

Safety Evaluation and Regulation of Chemicals. 1st Int. Conf., Boston, Mass., 1982, pp. 277–283 (Karger, Basel 1983)

# Differential Cytotoxicity Testing of Medical Device Materials

*M.P. Bear, D.S. Johnson, F.H. Schneider*

Bioassay Systems Corporation, Woburn, Mass., USA

Cell culture methods have been effectively used for evaluating the potential toxicity of medical device materials. Their cost effectiveness, sensitivity and rapidity recommend them as alternative methods to animal test systems. In addition to primary screening, cell culture methods can also be used for new material development programs. These programs require quantitative and sensitive tests capable of differentiating cytotoxicity among different materials. Through this type of analysis it is possible reliably to rank the cytotoxicity of materials and selectively to eliminate cytotoxic materials. Sensitive and quantitative cell culture systems may also be used for process control evaluation and comparative cytotoxicity testing on competitive products.

Historically, there have been two commonly used cell culture methods for screening the toxicity of medical device materials, the agar overlay (AO) [1] and MEM elution [2] tests. These test methods have been used to predict whether medical device materials would pass a rabbit intramuscular implant test, the animal model commonly used for evaluating the biocompatibility of medical device materials. However, the relative insensitivity and qualitative nature of these cell culture tests makes these methods ineffective for use in new material and process control evaluation studies.

A dual testing procedure for sensitive and quantitative cytotoxicity evaluation of medical device materials has been developed at Bioassay Systems. This test system employs the neutral red uptake (NRU) and the Chinese hamster ovary cell (CHO) clonal cytotoxicity tests. This combination of tests is referred to as a DNC analysis.

In this report, data will be presented which illustrate the differences in sensitivity among the AO, MEM elution, NRU and the CHO clonal cytotoxicity tests. Results will also be shown from three representative projects which illustrate the effectiveness of the DNC analysis for differential cytotoxicity evaluation of medical device materials.

## *Materials and Methods*

### *Cells*

Three different cell types were employed in this study: mouse L929 fibroblasts (Flow Laboratories), Flow 2000 human diploid fibroblasts (Flow Laboratories), and CHO cells (ATCC, CCL 60). Each of the three lines was maintained in exponential growth in monolayer culture.

### *Sample Preparation*

In this study, various different medical device materials were evaluated, including plastic resins, surgical latex, nonwoven materials, and fabrics. The resins were received as pellets; the other materials were received in sheets. Using aseptic technique, the sheets were cut into approximately 1-$cm^2$ pieces. For the AO test the materials were applied directly to the test plates without further processing. For the extraction tests, the subdivided materials were extracted in the appropriate cell culture medium (supplemented with 10% fetal calf serum) at a ratio of 60 $cm^2$/ 20 ml extraction medium. Extractions were conducted in a moist atmosphere of 95% air/5% $CO_2$ at 37 °C for 24 h. The extracts were collected and used directly in the direct exposure and NRU tests. However, for the CHO test, the extracts were filtered through a 0.20-$\mu$m membrane filter prior to testing. Membrane filtration of this extract is required since the length of the test (5–7 days) is greater than the half-life of the antibiotic used to ensure sterility.

### *Test Procedures*

*AO.* L929 cells in exponential growth were seeded into 20 × 100-mm Petri dishes at an inoculum density of $1 \times 10^6$ cells per dish and allowed to reach saturation density after 4 days in culture. At this time, the medium was aspirated from each plate and 12.5 ml of AO was added; 10 ml of a 0.04% solution of neutral red was added to each plate and incubated at 37 °C for 15 min and then removed by aspiration. The plates were incubated for at least 1 h and then examined for even staining. Using aseptic technique, the test and control samples were placed directly on the agar surface at the corners of a centrally located estimated 4-cm square. Each plate contained two test samples and positive and negative controls. Each test sample was tested on a minimum of two plates. The plates were incubated for 24 h in a moist atmosphere of 95% air/5% $CO_2$ at 37 °C. After incubation, the L929 monolayers were examined microscopically for zones of decolorization and cell lysis. The degree of cytotoxicity for each test and control sample was scored according to the dual scoring system of *Guess* et al. [1] and presented as a decolorization/ lysis ratio. Only sample(s) producing cell lysis are considered cytotoxic.

*Direct Exposure.* 1-ml aliquots of the test and control samples were added to duplicate cultures of confluent monolayers of Flow 2000 human diploid fibroblasts grown on glass coverslips in a 24-well plate. The cells were treated for 24 h in a moist atmosphere of 95% air/5% $CO_2$ at 37 °C. The cells were then fixed and stained with hematoxylin and observed microscopically for morphological signs of cytotoxicity as judged by cell lysis and vacuole formation. The degree of cytotoxicity was scored on a relative scale from 0 to 4 based on the percentage of affected cells.

*NRU.* Flow 2000 human diploid fibroblasts were seeded into 24-well plates at an inoculum density of $0.15 \times 10^6$ cells per well. After 24 h of incubation, the media were aspirated from the wells and 1.0-ml aliquots of the test and control samples were added to triplicate wells. The cells were treated for 48 h in a moist atmosphere of 95% air/5% $CO_2$ at 37 °C. After 48 h of treatment, 40 μl of a neutral red working solution (1 mg/ml) was added to each well and the cultures were incubated for 2 h, after which the culture medium was aspirated from each well. The cultures were gently rinsed three times with phosphate-buffered saline to remove excess stain. The intracellular neutral red was extracted with 1.0 ml of extraction buffer and measured spectrophotometrically at 540 nm. The percent neutral red uptake was based on the amount of neutral red extracted from the medium control cultures.

*Clonal Cytotoxicity Test.* CHO cells obtained from cultures in exponential growth were inoculated into 20 × 100 mm Petri dishes at a density of approximately 1,000 cells per plate. After 24 h of incubation, the medium was aspirated from the cultures and the undiluted test extract and four dilutions of the extract were added to duplicate plates. Cell culture medium was added to five plates as the negative control and five concentrations of cadmium chloride were added to an additional set of duplicate plates as the positive control. After 5–7 days of incubation in a moist atmosphere of 95% air/5% $CO_2$ at 37 °C, the cells were fixed, stained with 0.5% crystal violet, and the surviving colonies counted using an automatic colony counter. The cytotoxicity of the test and control samples was quantitated by determining the concentrations which inhibited colony survival by 50% ($EC_{50}$). The $EC_{50}$ values were derived graphically from concentration response curves.

*Dual NRU/CHO Clonal (DNC) Scoring System.* A scoring system was developed to rank the relative cytotoxicity of test materials evaluated in the DNC analysis. This scoring system is based on assigning toxicity points to the response obtained in each test. The cytotoxic response produced by a test material is scored as follows:

NRU: toxicity points = 100% – % uptake score;

$$\text{CHO clonal (CHO): toxicity points} = \frac{EC_{50}\,(\text{ref.}) \times 200}{EC_{50}\,(\text{test sample})};$$

total toxicity points = NRU points + CHO points.

More toxicity points are assigned to the response obtained in the CHO test because this test is more sensitive than the NRU test. Furthermore, CHO toxicity points are based on the toxicity of a reference material. In the studies reported here, the reference material was defined as the sample within each test group which exhibited the greatest degree of cytotoxicity.

## *Results*

Table I presents results from a comparative cytotoxicity study conducted on five different medical device materials. The objective of this study was to rank the relative degree of cytotoxicity of the five test materials. The cytotoxicity of each material was tested in four assays including the AO, MEM elution, NRU and the CHO clonal cytotoxicity tests. The results presented in table I show the differences in sensitivity among the four test systems. The AO test was the least sensitive of the four tests. It was unable to detect extractable cytotoxic substances from any of the five materials. Although there were

*Table I.* Project 1: comparison of four cytotoxicity tests

| Sample | AO (decol./lysis) | Direct exposure (lysis) | NRU, % | CHO clonal $EC_{50}$, % |
|---|---|---|---|---|
| Nylon towels | 1/0, 1/0<br>1/0, 1/0 | 0, 0 | 104 | 49 |
| Rice paper | 0/0, 0/0<br>0/0, 0/0 | 0, 0 | 119 | nontoxic |
| Cellulose nitrate | 0/0, 0/0<br>0/0, 0/0 | 0, 0 | 96 | nontoxic |
| 99% polypropylene | 2/0, 2/0<br>2/0, 2/0 | 4, 4 | 47 | 38 |
| 97% polypropylene | 2/0, 2/0<br>2/0, 2/0 | 3, 3 | 28 | 15 |

Nontoxic means that no inhibition in CHO colony survival was observed at all concentrations tested.

*Table II.* Project 1: cytotoxicity ranking of materials using the DNC analysis

| Sample | NRU | | CHO | | Total toxicity pts | Toxicity rank |
|---|---|---|---|---|---|---|
| | uptake, % | tox. pts | $EC_{50}$, % | tox. pts | | |
| Rice paper | 104 | 0 | nontoxic | 0 | 0 | 1 |
| Cellulose nitrate | 96 | 4 | nontoxic | 0 | 4 | 2 |
| Nylon towels | 104 | 0 | 49 | 62 | 62 | 3 |
| 99% polypropylene | 47 | 53 | 38 | 79 | 132 | 4 |
| 97% polypropylene | 28 | 72 | 15 | 200 | 272 | 5 |

$EC_{50}$ = Percent of test extract inhibiting colony survival by 50%. Total toxicity points = scoring system is defined in Materials and Methods. Nontoxic means that sample extract did not inhibit colony survival.

differences in the decolorization responses, none of the five materials produced cell lysis. The MEM elution and NRU tests were more sensitive than the AO test as indicated by the ability of these tests to detect cytotoxicity associated with the 97 and 99% polypropylenes. However, the CHO test was clearly the most sensitive of the four tests. Not only did it detect cytotoxicity of the polypropylenes, but it was the only test able to detect cytotoxicity associated with the nylon towels.

*Table III.* Project 2: comparison of three cytotoxicity tests

| Sample | AO (decol./lysis) | NRU, % | CHO clonal, $EC_{50}$, % |
|---|---|---|---|
| Absorbent nonwoven | | | |
| A | 0/0, 0/0<br>0/0, 0/0 | 113 | 77 |
| B | 0/0, 0/0<br>0/0, 0/0 | 107 | 73 |
| C | 1/0, 1/0<br>1/0, 1/0 | 105 | 66 |
| Surgical latex | | | |
| A | 0/0, 0/0<br>0/0, 0/0 | 121 | nontoxic |
| B | 2/0, 2/0<br>2/0, 2/0 | 20 | 5 |
| C | 2/1, 2/0<br>2/1, 2/1 | 2 | 1 |

Nontoxic means that no inhibition in CHO colony survival was observed at all concentrations tested.

Since the AO and MEM elution tests are qualitative in design, results from these tests cannot be used to rank the relative degree of cytotoxicity of the five materials. Ranking the materials can be accomplished using the quantitative dual NRU/CHO clonal analysis (DNC) as indicated in table II. This table illustrates the quantitative differences in cytotoxicity existing among the five test materials. Rice paper and cellulose nitrate were ranked as the least cytotoxic with total toxicity points of 0 and 4, respectively. The nylon towel material scored 62 total toxicity points and its cytotoxicity was differentiated from that of either the rice paper or the cellulose nitrate. Most importantly, the DNC analysis differentiated the cytotoxicity of the 97 and 99% polypropylenes; the 97% polypropylene being approximately twice as toxic as the 99% polypropylene. This unusual concentration-response relationship is probably independent of the polypropylene concentration, but associated with unidentified cytotoxic substances included in the material.

Tables III and IV also illustrate the differences in sensitivity among the AO, NRU and CHO tests and the usefulness of the DNC analysis for ranking materials according to their cytotoxicity. Two groups of three materials each were evaluated in this study including absorbent nonwoven materials and surgical latex. Neither the AO nor the NRU tests detected any cytotoxic extractables from the absorbent nonwoven materials (table III). However,

*Table IV.* Project 2: cytotoxicity ranking of materials using the DNC analysis

| Sample | NRU | | CHO | | Total toxicity pts | Toxicity rank |
|---|---|---|---|---|---|---|
| | uptake, % | tox. pts | $EC_{50}$, % | tox. pts | | |
| Absorbent nonwoven | | | | | | |
| A | 107 | 0 | 77 | 172 | 172 | 1 |
| B | 113 | 0 | 73 | 182 | 182 | 2 |
| C | 105 | 0 | 66 | 200 | 200 | 3 |
| Surgical latex | | | | | | |
| A | 121 | 0 | nontoxic | 0 | 0 | 1 |
| B | 20 | 80 | 5 | 39 | 119 | 2 |
| C | 2 | 98 | 1 | 200 | 298 | 3 |

Nontoxic means that sample extract did not inhibit colony survival.

the extracts prepared from the absorbent nonwovens were cytotoxic in the CHO test. Table III also shows that the NRU and CHO tests were considerably more sensitive to cytotoxic extractables from samples B and C of the surgical latex than was the AO test. Based on their $EC_{50}$ values, the CHO test was more sensitive to cytotoxic extractables than was the NRU test.

By using the DNC scoring system, the relative cytotoxicity of the test materials could be reliably ranked (table IV). Of the surgical latex materials, sample A was noncytotoxic in all three tests and received a score of 0. The cytotoxicities of sample B and C were clearly differentiated, with sample C being approximately 2.5 times more toxic than sample B. This differentiation in cytotoxicity between samples B and C could not be obtained using the AO test. The cytotoxicity of the absorbent nonwovens could be differentiated based upon the results from the CHO test, with their relative ranking indicated in table IV. However, for these samples, additional data are required to determine whether these differences in cytotoxicity are statistically significant.

Table V presents the results from a DNC cytotoxicity analysis conducted on forty samples of plastic resins considered for use in the development of a new medical device. The results from six of the forty samples are presented in this table to illustrate the effectiveness of the DNC analysis for differentiating cytotoxicity among the test materials. The six materials exhibited considerable differences in their cytotoxicity and based on the quantitation of cytotoxicity, candidate materials could be selectively eliminated from future

*Table V.* Project 3: cytotoxicity ranking of plastic resins using the DNC analysis

| Sample code | NRU | | CHO | | Total toxicity pts | Toxicity rank |
|---|---|---|---|---|---|---|
| | uptake % | tox. pts | $EC_{50}$ % | tox. pts | | |
| A | 104 | 0 | nontoxic | 0 | 0 | 1 |
| B | 81 | 19 | 89 | 62 | 81 | 10 |
| C | 71 | 29 | 85 | 65 | 94 | 15 |
| D | 43 | 57 | 90 | 61 | 118 | 20 |
| E | 25 | 75 | 71 | 78 | 153 | 30 |
| F | 40 | 60 | 28 | 200 | 260 | 40 |

studies. The distribution of cytotoxicity among the forty materials was as follows: very toxic (> 150 points) = 20%; moderately toxic (100–150 points) = 25%; slightly toxic (26–100 points) = 25%, and not toxic (< 25 points) = 30%.

## *Conclusion*

We have developed a dual test procedure using the NRU and the CHO clonal cytotoxicity tests for quantitative and sensitive cytotoxicity evaluation of medical device materials. The results reported in this paper illustrate the effectivness of this procedure for reliably ranking and differentiating cytotoxicity among different types of medical device materials. It provides an effective method applicable to new material development and process control evaluation. This paper also illustrates the relative insensitivity and qualitative nature of the AO and MEM elution tests which render these tests ineffective for this type of quantitative in vitro cytotoxicity analysis.

## *References*

1 Guess, W.L.; Rosenbluth, S.A.; Schmidt, B.; Autian, J.: Agar diffusion method for toxicity screening of plastics on cultured cell monolayers. J. pharm. Sci. *54:* 156 (1965).

2 Wilsnack, R.E.; Meyer, F.J.; Smith, J.G.: Human cell culture toxicity of medical devices and correlation to animal test. Biomater. med. Devices artif. Organs *1:* 543–562 (1973).

Dr. M.P. Bear, Bioassay Systems Corporation, 225 Wildwood Avenue,
Woburn, MA 01801 (USA)

# Subject Index